William H. Ashmead

Classification of the bees, or the superfamily Apoidea

William H. Ashmead

Classification of the bees, or the superfamily Apoidea

ISBN/EAN: 9783741185632

Manufactured in Europe, USA, Canada, Australia, Japa

Cover: Foto ©Andreas Hilbeck / pixelio.de

Manufactured and distributed by brebook publishing software
(www.brebook.com)

William H. Ashmead

Classification of the bees, or the superfamily Apoidea

CLASSIFICATION OF THE BEES, OR THE SUPERFAMILY APOIDEA.

BY WILLIAM H. ASHMEAD.

In the Journal of the New York Entomological Society, for March, 1899, I separated the Hymenoptera into ten superfamilies. The first of these or the Apoidea comprises the bees, among which, especially among the social bees, are to be found probably the highest or most specialized types in the order; hence my reason for beginning the classification of the Hymenoptera with these insects.

Our own bees, and indeed the bees of most countries, except those of the European fauna, are but little studied and very imperfectly known.

Thomas Say, Frederick Smith, Ezra T. Cresson, Charles Robertson, Abbe Provancher, Wm. J. Fox, T. D. A. Cockerell and a few others have done much towards making our species known, but there is still much to be done before we shall gain a knowledge of the immense number of species found in our vast country. Our study of them is just begun.

Prof. T. D. A. Cockerell's recent work on the bee fauna of the arid regions of New Mexico, Arizona, etc., illustrates well how little we really know of our bee fauna and what may be accomplished by one energetic student in the way of turning up new or undescribed species in a comparative limited area.

What Prof. Cockerell has done with the bee fauna of his part of the country could, I feel sure, be duplicated by energetic collectors and students in other parts of the country, since, I believe, we know scarcely twenty per cent of our indigenous species. There is, therefore, an immense unexplored field, offering the best opportunity for original work and discoveries, still opened to the student who will take up the study of our bees.

It is earnestly hoped, therefore, that the early publication of these tables will stimulate, aid and encourage our younger students to take up and study these neglected insects.

Before proceeding with my tables I desire briefly to call special attention to two most valuable works, treating upon the European bee fauna, which have appeared lately, and upon which much of my own work is based, viz.,

(1) Apidæ Europææ per genera, species et varietates dispositæ atque descriptæ a Dr. H. L. O. Schmiedeknecht. 1882–86.

(2) Die Bienen Europa's (Apidæ Europææ) nach ihren Gattungen, Arten und Varietäten auf vergleichend morphologisch-biologischer Grundlage bearbeitet von Heinrich Friese, 1893 97.

These books are most valuable and ought to be in the hands of all students who contemplate taking up the study of the bees. I have found them almost invaluable in my studies on the structure and affinities of this large complex group of insects.

It will also be observed that I have drawn quite largely upon Schmiedeknecht's work for ideas on the classification of these insects. In fact, minus a few slight changes, I have followed his ideas in extenso, as regards families, as may be readily seen by a comparison of my arrangement with his.

Dr. Schmiedeknecht, after giving an excellent historical review and resumé of the various schemes proposed for classifying these insects, during the past century or more, on page 11 * of his work, proposed the following arrangement:—

Section I.—Apidæ sociales.
Section II.—Apidæ solitariæ, constructing cells.
Section III.—Apidæ parasiticæ.

Section I.

This section he has divided into two distinct families—*A* and *B*.

A.—Fam. I.   Apidæ sens. str.   Permanently social.—*Apis.*
B.—Fam. II.   Bombidæ.   Social but once a year.- *Bombus.*

Section II.

In this section Schmiedeknecht distinguishes three principal groups:

A.—Podilegidæ (Scopulipedes, leg collectors).
    a. Crurilegidæ (Tibia collectors).
    b. Femorilegidæ (Femur collectors).
B.—Gastrilegidæ (Dasygastræ, belly collectors).
C.—Pseudoparasitæ, *i. e.*, not furnished with an apparatus for collecting pollen, but yet not living parasitically.

Aa,—contains two families:
    Fam. III.   Anthophoridæ.—*Anthophora, Habropoda, Saropoda, Macrocera, Plistotrichia, Eucera, Meliturga, Systropha.*
    Fam. IV.   Melittidæ.— *Cilissa, Macropis.*
Ab,—contains three families:
    Fam. V.   Xylocopidæ.—*Xylocopa, Ceratina.*
    Fam. VI.   Panurgidæ.—*Panurgus, Dasypoda, Panurginus, Camptobæam, Dufourea, Biareolina, Rhophites, Rhophitoides, Halictoides.*
    Fam. VII.   Andrenidæ.—*Andrena, Colletes, Nomia, Nomioides.*

B,—contains but a single family :

Fam. VIII. Megachilidæ.—*Megachile, Chalicodoma, Lithurgus, Trachusa, Osmia, Heriades, Trypetes, Chelostoma, Anthidium.*

C,—contains two families :

Fam. IX. Sphecodidæ.—*Sphecodes.*

Fam. X. Prosopidæ.—*Prosopis.*

SECTION III.

This section he has divided into two groups :

A.—Inquilines or Commensales, *i. e.,* parasitic bees living in the nests of social species.

Fam. XI. Psithyridæ.—*Psithyrus (= Apathus).*

B.—True parasitic bees.

Fam. XII. Melectidæ.—*Melecta, Crocisa, Nomada, Epeolus, Epeoloides, Biastes, Pasites, Melittoxena, Ammobates, Ammobatoides.*

Fam. XIII. Stelididæ.—*Stelis, Cœlioxys, Dioxys.*

This arrangement of Dr. Schmiedeknecht's, and his separation of the old Apidæ into many well defined and readily recognizable families, I consider a decided improvement over all schemes of arrangement proposed by those who have preceded him, and will, I feel sure, in time meet with the hearty support and approval of all students of the bees.

Dr. Henry Friese, in his work, which is substantially a continuation of Schmiedeknecht's, follows closely the latter's arrangement, except that he treats his families as subfamilies, and has recognized 14 subfamilies in place of 13 families. This additional subfamily is made by separating from Schmiedeknecht's family Stelididæ, the genus *Cœlioxys* and others, which he calls the Cœlioxinæ.

Dr. Friese reverses the order of arrangement, proposed by Schmiedeknecht, and begins with the Prosopinæ or less specialized bees, and ends with the Apinæ, or the most highly specialized.

It is so nearly identical with Schmiedeknecht's arrangement as not to require a repetition here.

In the following pages it will be seen that I have in the main recognized most of the families, as proposed by Schmiedeknecht, but instead of thirteen I have fourteen, not entirely agreeing with either Schmiedeknecht nor Friese. How this is done may be readily discerned by an examination of my table of families, which is to follow.

Three or four of Schmiedeknecht's families I consider unnatural, or at least not equivalent to family rank, or to the rank of his other families, and two are suppressed. His family Sphecodidæ I have merged as a subfamily with the Andrenidæ, since it agrees in every respect with many genera in this family, *Halictus* and allies, except that the pollen brush or flocculus is wanting on the hind femora, and thin and sparse, or scarcely noticeable on the hind tibiæ and tarsi.

It is not quite exact to say that the flocculus is wanting on hind legs in the Sphecodinæ, for it is really present, although much reduced, but yet sufficiently developed to retain pollen.

His family Melittidæ, containing only two genera, is also suppressed: *Cilissa* being placed with the Andrenidæ, *Macropis* with the Panurgidæ.

I also consider his family Melectidæ—my family Nomadidæ, in part (= subfamily Nomadinæ Friese)—a composite one; many of the genera he has placed in it, such as *Pasites, Melittoxena, Ammobates*, etc., not belonging to it, but, according to my views, should be placed with the Stelididæ. Their relationship to the Nomadidæ, if it ever existed, must be certainly very distant.

For the genus *Ceratina*, placed by Schmiedeknecht in his family Xylocopidæ, I have erected the family Ceratinidæ, since it seems to me to present scarcely any character in common.

The genus *Colletes*, Schmiedeknecht placed with the Andrenidæ. The Andrenidæ, as now restricted, will not permit this arrangement, since the mouth parts in *Colletes* are too entirely different from those of the genuine Andrenidæ. It is clearly related to the Prosopidæ, where Bingham (Fauna India Hymenoptera) and Friese have placed it.

It differs, however, in several particulars from *Prosopis* and allies. For instance, the front wings have three cubital cells, the head and thorax are densely pubescent, while the hind legs in the ♀ have a dense scopa, characters of great taxonomic importance (not possessed by *Prosopis*), and which seem to me to forbid its retention with the Prosopidæ.

There are several genera closely related to *Colletes*, having all the above characters in common, and for these I have retained the family name Colletidæ, first proposed by Col. Bingham.

In the arrangement of the families and genera in this work, I have attempted to bring together or in juxtaposition, such as have had seemingly a common origin or ancestry, or exhibit strong affinities. For I believe with Dr. Friese and others, that the commensals and parasitic bees are nothing more than offshoots from other bees, just as I believe and have already published, that the Ichneumonids are offshoots from plant-feeding wasps, Siricidæ, etc.; that the inquilinous and parasitic Cynipids are offshoots from the plant-feeding Cynipids, and that the parasitic Chacidids are offshoots from the plant-feeding Chalcidids, Agaonidæ, Eurytomidæ, etc.

This same kind of development or evolution is also found existing among the wood, sand and digger wasps: Masaridæ and Chrysididæ originated from the Vespidæ and Eumenidæ, while *Ceropales*, a genus in the Pompilidæ, living parasitically in the nests of other species of Pompilids, is an offshoot from *Pompilus*.

Among the bees, *Psithyrus* (= *Apathus*) is clearly an offshoot from *Bombus*, as Friese has shown.

The Nomadidæ evidently came directly from the Anthophoridæ and other pollen-gathering bees, while the Stelididæ for the most part had their origin among the Megachilidæ—the subfamily Stelidinæ coming from the Anthidiinæ, the Cœlioxinæ from the Megachilinæ, etc.

The Panurgidæ, or at least part of them, are of quite recent origin, some genera being evidently only recently evolved from *Andrena* and *Halictus*, while others seem to have come from the Anthophoridæ.

In our classificatory work I believe this law of evolution or development (for certainly evolution is no longer a theory, but a demonstrated law of the universe), should be borne in mind, and, as far as possible, the origin and affinities of the complexes, such as families, groups, genera, etc., should be interpreted and shown in our tables.

With this object in view, in the present work, therefore, I have not recognized the three sections of bees defined by Schmiedeknecht, since, to do so, would separate widely closely allied families; but instead I have interpolated, in my tables, the inquilinous and parasitic bees among the true honey-making bees, in order to demonstrate, what I conceive to be, their true affinities and relationship to each other.

The families into which the *Apoidea* are now divided may be separated by the use of the following table:

Superfamily I. Apoidea.

Table of Families.

Labium or tongue flattened, most frequently shorter than the mentum, rarely very much longer (some Panurgidæ); basal joints of labial palpi cylindrical, the first joint sometimes very elongate or thickened, but still neither flattened nor unlike the following................................6.

Labium or tongue very elongate, slender and always longer than the mentum; two basal joints of labial palpi very elongate, compressed, valvate and very unlike the following, which are minute, the third or fourth joint uniting with the second a little before its apex.

Hind tibiæ *with* two apical spurs....................................2.

Hind tibiæ *without* apical spurs.

Sexes three, ♀, ☿, ♂; workers with corbiculæ, ♀ *without;* maxillary palpi very short, 1-jointed; labial palpi 4-jointed, with the joints very unequal, the first two long, valvately compressed....Fam. I. APIDÆ.

2. First submarginal cell not, or rarely, divided by a delicate, oblique nervure; if at all present, incomplete or indicated by a hyaline streak or nervure; sexes two, ♀, ♂; hind tibiæ in ♀ outwardly convex or rounded, never concave; no corbiculæ; basal joint of hind tarsi in ♀ not forcipate at base; malar space, except in Psithyridæ, wanting or indistict, never very large....................................3.

First submarginal cell most frequently divided by a distinct, but delicate, oblique nervure, rarely indistinct; hind tibiæ and metatarsi in ♀ strongly dilated, outwardly concave (in a single case only convex, but this has a channel along the hind margin *Aglæ*); metatarsus in ♀ forcipate at base; malar space large, distinct.

Labrum transverse, subtrapezoidal, the clypeus not carinate; body densely hairy; scutellum semicircular, rounded off posteriorly and not projecting over the metanotum; sexes three, ♀, ☿, ♂; ♀ and ☿ *with* corbiculæ, and a dense polleniferous scopa on hind tibiæ and tarsi; labial palpi 4-jointed; maxillary palpi short, 2-jointed; tongue not extending beyond apex of thorax..............Fam. II. BOMBIDÆ.

Labrum large, subquadrangular, the clypeus, and most frequently the labrum also, carinate; body most frequently metallic, bare or only slightly pubescent, rarely very densely pubescent *Eulema;* scutellum large, quadrangular, projecting over the metanotum, the axillæ very small; sexes two, ♀, ♂, the ♀ with corbiculæ, but with the polleniferous scopa on hind tibiæ and tarsi very sparse and thin, and confined to the lateral edges; labial palpi 2-jointed; maxillary palpi 1-jointed; tongue reaching to or beyond the middle of abdomen.

Fam. III. EUGLOSSIDÆ.

3. Front wings with two submarginal cells....................................5.

Front wings with three submarginal cells.

Eyes extending to or nearly to the base of the mandibles, the malar space wanting or at most not longer than the pedicel....................4.

Eyes not nearly extending to the base of the mandibles, the malar space large, distinct, longer than the pedicel and first joint of flagellum united.

Marginal cell very long, as long or longer than the three submarginal cells united; body rather densely pubescent; abdomen broadly oval or oblong, flat beneath, convex above; ♀ without a polleniferous scopa; ♂ with the eyes frequently strongly convergent above, the genitalia, squama and lacinia always membranous...Fam. IV. PSITHYRIDÆ.

4. Labrum large, free, convex or inflexed.

Marginal cell not especially long or narrow, rarely longer than the first two submarginal cells united.

♀ *with* a dense polleniferous scopa on hind tibiæ and tarsi; body clothed with a dense pubescence: maxillary palpi 2–6 jointed.
Fam. V. ANTHOPHORIDÆ.

♀ *without* a polleniferous scopa or at most with a thin, sparse flocculus on hind tibiæ; body most frequently bare or nearly; the pubescence, if any, short and sparse, rarely densely pubescent; species often metallic or rufous and black, the abdomen most frequently ornate, with white or yellow maculæ or bands................Fam. VI. NOMADIDÆ.

Marginal cell long and narrow, usually as long or longer than the three submarginal cells united.

Hind tibiæ and tarsi with a sparse pubescense, but *without* a distinct scopa : maxillary palpi usually 6-jointed ; body metallic or submetallic, nearly bare; abdomen elongate, subcylindrical, the segments more or less constricted at sutures. Small species.....Fam. VII. CERATINIDÆ.

Hind tibiæ and tarsi with a dense scopa; maxillary palpi 5–6-jointed, rarely wanting; thorax more or less densely pubescent, at least laterally; abdomen not elongate, with a scopa beneath ; eyes in ⚥ often strongly convergent above ....Fam. VIII. XYLOCOPIDÆ.

5. Labrum large and free, uncovered: maxillary palpi 4, 5 or 6-jointed (rarely wanting *Oxæa*): body densely pubescent, the hind legs with a dense scopa; ventral scopa present.

Marginal cell neither long nor narrow..Fam. V. ANTHOPHORIDÆ (pars).
Marginal cell very long and narrow....Fam. VIII. XYLOCOPIDÆ (pars).

Labrum not large and free, most frequently entirely covered by the clypeus (Megachilidæ), or, if somewhat visible, then strongly inflexed (*Stelididæ*).

Abdomen in ♀ *with* a ventral scopa; labrum entirely covered by the clypeus.
Fam. IX. MEGACHILIDÆ.

Abdomen in ♀ *without* a ventral scopa; labrum more or less visible, not entirely covered by the clypeus, strongly inflexed.
Fam. X. STELIDIDÆ.

6. Labium or tongue short, broad, obtuse or emarginate at apex, never acute medially; hind femora with or without a distinct golden brush or flocculus.............   ....7.

Labium or tongue long or short, but always acute medially at apex ; hind femora always with a pollen brush or flocculus, rarely very thin and sparse.

Front wings with two submarginal cells: labium or tongue long or short, usually, however, narrowed and longer than the mentum; labrum rather large, distinct, not covered by the clypeus, but most frequently inflexed ....Fam. XI. PANURGIDÆ.

Front wings with three submarginal cells; labium or tongue shorter or not longer than the mentum, triangular, not narrowed, rarely long; labrum not free, more or less hidden by the clypeus, or with basal processes always visible....................Fam. XII. ANDRENIDÆ.

7. Front wings with three submarginal cell; head and thorax more or less clothed with a dense pubescence ; second recurrent nervure often more or less sinuate : lingua at apex rather deeply triangularly emarginate ; hind femora in ♀ with a pollen brush or flocculus.
Fam. XIII. COLLETIDÆ.

Front wings with two submarginal cells; head and thorax bare or nearly;
second recurrent nervure always straight; lingua very short and
broad, shallowly or very obtusely triangularly emarginate at apex;
hind femora without a pollen brush or flocculus.

Fam. XIV. PROSOPIDÆ.

Family I. APIDÆ.

To this family, as here restricted, belong all the genuine honey-
making social bees, living in large communities and consisting of
three sexes, females (or queens), workers and males. Here belong
the common hive-bee, the domesticated bees of various countries
and the stingless bees of subtropical and tropical regions.

The family is at once distinguished from all the other families of
bees by the total absence of apical spurs on the hind tibiae, by the
venation of the front wings, and by the workers being furnished
with corbiculæ. This last character is absent in all other bees,
except in the Bombidæ and Euglossidæ, which, however, are large,
robust bees, with two apical spurs on the hind tibiae, while the venation of the front wings is quite different.

Two subfamilies have been recognized, distinguished as follows:

Table of Subfamilies.

Front wings with two (rarely three) distinct cubital cells, the marginal cell
lanceolate, slightly open at apex; stigma lanceolate or narrow ovate;
eyes bare, extending to base of mandibles; ♀ and ☿ without a sting,
but both with corbiculæ.................Subfamily I. MELIPONINÆ.
Front wings with three distinct cubital cells, the marginal cell rounded and
closed at apex; eyes hairy, not extending to base of mandibles; ♀ and
☿ with a sting, ☿ only with corbiculæ; eyes in ♂ holoptic.
Subfamily II. APINÆ.

Subfamily I. MELIPONINÆ.

(The Stingless Honey-bees).

This subfamily is readily distinguished from the common honey or
hive-bees by the venation of the front wings, the bare eyes, which
extend to the base of the mandibles, the ♀ and ☿ being without a
sting, and by the *simple* not cleft claws.

Three genera have been recognized, only two of which are known
to me, *Melipona* and *Trigona*. Some authors would unite both
under the older name *Melipona*, but since they are readily separable,
I believe it best to retain both genera.

Table of Genera.

Thorax rather densely pubescent; mandibles broad, without teeth, the apex broad and blunt; stigma in front wings lanceolate.

Abdomen not small, convex above, the venter rounded, not carinate, with a distinct scopa.... ....**Melipona** Illiger.

Thorax and abdomen bare or almost bare; mandibles broad, but with teeth, or at least with one or two small teeth at inner apical angles; stigma in front wings subovate or ovate-lanceolate.

Abdomen small, short, triangular, carinate beneath........**Trigona** Jurine.

Abdomen elongate, almost quadrangular, the dorsum forming an obtuse angle.
Tetragona Seville et. Lepel.

Subfamily II. APINÆ.

(The Hive or true Honey-bees).

To this family belong the true honey-bees, the hive bee and the various domesticated bees of different countries—the Italian bee, the Egyptian bee, the bee of India, etc. They differ in several ways from the Meliponinæ; the ♀ and ☿ always possess a sting, the eyes are hairy and do not quite extend to the base of the mandibles, while the front wings have three distinct cubital cells, and the marginal cell is long and always closed and rounded at apex, quite different from that of the Meliponinæ. They also differ in having cleft, not simple, claws.

This family is represented by a single genus, *Apis* Linné, readily recognized by the characters given in my table of subfamilies.

Family II. BOMBIDÆ.

(The Bumble or Humble Bees).

The bees belonging to this family comprise median to large sized robust bees, clothed with a dense, more or less velvety-like pubescence, and known to us under the name of bumble or humble bees. In their habits they agree with the Apidæ, being social and living in large communities, each species being composed of three kinds of individuals—males, females and workers, the latter being sometimes called neuters or nurses

Only a single genus is known, distinguished as follows:

♀ and ☿ with the posterior tibiæ dorsally depressed, polished and furnished with corbiculæ, posterior tarsi with the first joint angulated above forming a forcipate hook externally.

☿ with the posterior tibiæ above more or less shining, somewhat concave; genitalia, squama and lacinia corneous.................**Bombus** Fabr.

Family III. EUGLOSSIDÆ.

This family is erected to contain two genera of bees found in Mexico, Central and South America, viz., *Euglossa* and *Eulema*, usually placed with the Anthophoridæ, but which clearly, on account of the character of the hind legs, are not at all related to them, but show a closer affinity with *Bombus*, *Apis* and *Melipona*.

Table of Genera.

Marginal cell at apex narrowly rounded, always separated from the costa; second and third submarginal cells each receiving a recurrent nervure, cubital cells, along the cubitus, nearly equal, or the second is sometimes the shortest or smallest.

  Third cubital cell, along the cubitus, fully as long or distinctly longer than either the first or second; thorax and abdomen clothed with a short dense pubescence..2.

  Third cubital cell, along the cubitus, scarcely as long as the first, the second clearly smaller than either the first or third; body strongly metallic bare or nearly.

   Labrum and clypeus tricarinate; scutellum large, semicircular, with a median depressed line posteriorly, which, in ♀, is frequently filled with a cleft of black hairs.............................**Euglossa** Latr.

2. Labrum and clypeus tricarinate; scutellum large, quadrangular, flat, the lateral margins slightly reflexed, the ♀ usually with a tuft of black pubescence medially, as in *Euglossa*; maxillary palpi 1-jointed, compressed.............................**Eulema** Lepel.

Family IV. PSITHYRIDÆ.

(The Falsa Bumblee-Bee).

This family is monogenetic and comprises *Bombus*-like parasitic bees, easily and often confused with the genuine bumble or humble bees, their external structural characters being almost essentially the same. The following characters will, however, readily distinguish them:

♀ *without* corbiculæ, with the hind tibiæ dorsally convex and densely pilose; hind tarsi not forcipate at base; anus inflexed; ♂ with the hind tibiæ equally pilose, the genitalia, squama and lacinia always membranous ....**Psithyrus** Lepeletier.

Family V. ANTHOPHORIDÆ.

This is probably the most extensive family among the bees, and is found widely distributed over the entire globe. Unlike the honeybees and the bumble bees, all the species are solitary in their habits and consist of but two sexes—a ♂ and a ♀.

The following table will enable the student to determine the genera at present recognized:

Table of Genera.

Front wings with two cubital cells..20.

Front wings with three cubital cells.

  Cubital cells, along the cubitus, with one or another of the cells much longer
    or much shorter than another...3.

  Cubital cells, along the cubitus, more nearly equal, none very much longer or
    shorter than another; the first is most frequently the longest, or the
    second or third is sometimes the longest...................................6.

3. Third submarginal cell, along the cubitus, longer than the first, or at least of
    an equal length..1.

  Third submarginal cell, along the cubitus, equal to the first or very much
    shorter..5.

4. Second cubital cell, along the cubitus, distinctly shorter than the first or third,
    the third nearly as long as the first, or a little longer and much nar-
    rowed above, rarely are the first and second subequal, the second being
    most frequently much shorter than either the first or third, quadran-
    gular or wider than long...17.

  Second cubital cell, along the cubitus, somewhat longer than either the first
    or third; first discoidal cell about equal in length with the marginal
    cell; clypeus smooth; first recurrent nervure received by the second
    at or a little before its middle.

  Thorax clothed with a dense pubescence, the abdomen more or less bare,
    except at base; marginal cell rather short, obtuse or rounded, the third
    cubital cell much narrowed above; hind legs in ♀ with a long, dense
    scopa; mandibles 3-4 dentate.....................**Centris** Fabr.

5. Third cubital cell much narrowed above, the third transverse cubitus angu-
    lated or strongly curved inwardly before uniting with the radius..10.

  Third submarginal cell not or scarcely narrowed above, the third transverse
    cubitus nearly straight or only slightly curved outwardly; the third
    submarginal cell along the radius, therefore, as long as, or nearly as
    long as, along the cubitus.

  Third cubital cell, along the cubitus, never longer than the first, most fre-
    quently smaller or shorter, quadrate or nearly.....................6.

6. Third cubital cell not nearly quadrate, much narrowed above............13.

  Third cubital cell quadrate or nearly, never much narrowed above; first cubi-
    tal cell, along the cubitus, a little longer than the second or distinctly
    longer.

  Second cubital cell quadrate or nearly, distinctly shorter and smaller than
    the third, or then wider (higher) than long, or trapezoidal; clypeus in
    ♂ yellow; maxillary palpi 6-jointed... ....9.

  Second and third cubital cells, along the cubitus, equal or very nearly, fully
    as long or longer than wide.

  Clypeus in ♀ black, in ♂ yellow; maxillary palpi 5 or 6-jointed; abdo-
    men sometimes with pale fasciae; first recurrent nervure received by
    the second near or a little beyond its middle; first discoidal cell much
    longer than the marginal...7.

  Clypeus in both sexes yellow; maxillary palpi 4-jointed; abdomen with
    yellow fasciae or narrowly fasciate with white hairs.

                                              Saropoda Latreille.

7. Mandibles at apex bluntly rounded, truncate, or at most bidentate; labial palpi 4-jointed...8.

Mandibles in both sexes tridentate; labial palpi 4-jointed; first recurrent received by the second cubital cell a little beyond its middle, the second cubital slightly narrowed above............**Clisodon** Patton.

8. First recurrent nervure received by the second cubital cell at or a little beyond its middle (rarely a little before); front coxæ in ♂ normal, unarmed.

Abdomen black, sometimes banded or fasciate: maxillary palpi 6-jointed.

Malar space more or less distinct, but narrowed, not or scarcely as long as the second flagellum joint; transverse median nervure straight; first recurrent nervure received by the second cubital cell a little beyond its middle; abdomen without distinct fasciæ, bare, or with the segments 1-4 densely pubescent.

Anthophora Latreille = *Podalarius* Latr.

Malar space wanting, the eyes extending fully to the base of the mandibles.

Dorsal abdominal segments with large, white, hairy maculæ laterally.

Paramegilla Friese.

Dorsal abdominal segments *without* such maculæ, apically banded with ivory-white bands, or with pubescent fasciæ, or the whole abdomen is quite densely pubescent...........................**Amegilla** Friese.

Abdomen yellow or ferruginous, banded or maculate with black; maxillary palpi 5 or 6-jointed.

Abdomen ferruginous, the dorsal segments 1-4 with an oblong black spot on each side; maxillary palpi 6-jointed; mandibles dentate.

Lagobata Smith.

Abdomen yellow, banded with black; maxillary palpi 5-jointed; claws cleft.................. ....**Euthyglossa** Radoszk.

First recurrent nervure interstitial with the second transverse cubitus; front coxæ in ♂ armed with a long spine, the basal joint of the anterior and posterior tarsi much dilated; maxillary palpi long, 6-jointed, the first joint very short, the second the longest, as long as 4-6 united.

Habropoda Smith

9. First recurrent nervure almost interstitial with the second transverse cubitus: basal joint of hind tarsi in ♂ normal.

Second cubital cell almost quadrate, distinctly shorter than the third, the latter along the radius being about one-third shorter than along the cubitus; abdomen *without* fasciæ.......**Emphoropsis** Ashm. n. g.

First recurrent nervure received by the second cubital near its middle, the second recurrent received by the third cubital cell, either near its middle or near its apex; basal joint of hind tarsi in ♂ long, curved.

Second cubital cell trapezoidal, wider (higher) than long, but about twice as long along the cubitus as along the radius; abdomen not fasciate.

Emphor Patton.

10. Stigma distinctly developed, although never very long, but at least twice as long as wide..22.

Stigma very small, short, not longer than wide or subobsolete, never well developed (except in *Diadasia*); radial cell at apex remote from costa.

Submedian cell never much shorter than the median, equal or very nearly..11.

Submedian cell much shorter than the median..........................15.

11. First discoidal cell fully as long or distinctly longer than the marginal cell.
Basal joint of hind tarsi in ♂ normal, not curved, in ♀ produced at apex beyond the insertion of the second joint; flagellum not depressed. . 12.
Basal joint of hind tarsi in ♂ very long and curved or simply bent; flagellum in ♀ subdepressed; first recurrent nervure joining the second cubital cell near its middle, the second recurrent received by the third, either near its middle or beyond near its apex
Second cubital cell trapezoidal wider (higher) than long, but twice as long along the cubitus as along the radius; abdomen not fasciate.

Emphor Patton

12. Head rather large, the ocelli arranged in a transverse line; mandibles at apex bidentate; abdomen in ♀ 5-segmented, in ♂ 6-segmented; hind tibiae without a knee plate; claws bifid. . . . . **Moncea** Latr.
Head normal, the ocelli arranged in an obtuse triangle, or on a slight curved line; mandibles at apex simple or bidentate.
Second cubital cell, along the cubitus, distinctly shorter than the first or third. . . . . 14.
Second cubital cell, along the cubitus, as long or very nearly as long as the first, both being considerable longer than third . . . . 16.

13. Cubital cells, along the cubitus, about equal or subequal, the third along the radius only half as long as along the cubitus.
Second cubital cell a little larger than either the first or third, and along the radius longer than along the cubitus; first recurrent nervure received by the second before the middle; mandibles 3-4 dentate.

Centris Fabr.

Second cubital cell not larger than the first or second, usually a little shorter, and along the radius not longer than along the cubitus, usually shorter; first recurrent nervure received by the second cubital cell far beyond the middle, or interstitial with the second transverse cubitus; mandibles edentate, or at most bidentate at apex.
Abdomen in ♀ with 5 dorsal segments, not fasciate; antennæ filiform; claws simple. . . . . **Epicolpus** Spin.
Abdomen in ♀ with 6 dorsal segments, fasciate or subfasciate; claws cleft; maxillary palpi 5 jointed; clypeus in ♂ yellow, the antennæ very long. . . . . **Tetralonia** Spin.

14. Abdomen in ♀ with 6, in ♂ with 7 dorsal segments; antennæ in ♀ not long, subcompressed at apex, in ♂ very long; hind tibial spurs normal; claws cleft.
Maxillary palpi 6-jointed; second cubital cell quadrate or nearly, scarcely narrowed above, the first recurrent nervure received at its apical fourth. . . . . **Synhalonia** Patton.
Maxillary palpi 5-jointed.
First recurrent nervure received by the second cubital cell at its extreme apex, or very nearly interstitial with the second transverse cubitus, the second cubital cell longer than wide; thorax clothed with a whitish or cinereous pubescence; abdomen fasciate.

Tetraloniella Ashm. n. g. (Type *T. graga* Everm.).

First recurrent nervure received by the second cubital cell only a little beyond its middle, or at most at its apical third, the second cubital cell quadrate or a little wider (higher) than long; thorax clothed with a fulvous pubescence. . . . . **Xenoglossa** Smith

Maxillary palpi 4-jointed, the first and last joints very short, minute, scarcely
   longer than thick ; second cubital cell quadrangular, longer than wide,
   the first recurrent received at about its apical fourth.
                                                        Melissodes Latr. (pars).
15. Marginal cell at apex broadly truncate, with an appendage.

First cubital cell, along the cubitus, not longer than the second, the latter
   somewhat narrowed above, and receiving the first recurrent nervure
   at its extreme apex, or just before the second transverse cubitus ; third
   cubital cell a little the longest cell ; first discoidal cell shorter than the
   marginal ; dorsal abdominal segments broadly and sharply depressed
   at apex, the depressed portion being differently sculptured from the
   basal portion ; no fasciae on segments ; pygidial plate subtriangular,
   rounded at apex ; claws in ♀ simple, in ♂ cleft ; clypeus yellow in
   both sexes; mandibles bidentate ; eyes in ♂ convergent above, the
   face narrowed ; flagellum subclavate, the first joint of flagellum very
   long. . . . .**Meliturga** Latr.
First cubital cell, along the cubitus, longer than the second, the latter quad-
   rate, receiving the first recurrent nervure at its middle ; legs densely
   hairy, the middle tibial spur simple ; knee plate on hind tibiae want-
   ing ; claws cleft. . . . .**Melitoma** Latr.
16. Second cubital cell wider (higher) than long. . . . . .19.
Second cubital cell as long or longer than wide.

First and second cubital cells, along the cubitus, of an equal length or
   nearly, the third distinctly shorter and smaller than the others ; first
   recurrent nervure received by the second cubital cell a little beyond
   its middle ; clypeus prominent, bicarinate, the carinae convergent posteriorly ; mandibles at apex tridentate; maxillary palpi very short,
   2-jointed, the first joint short.
                               Epicharis Klug ? = *Melissoptila* Holmb.
17. Clypeus in ♂ not pale or yellow, at most and very rarely, with a pale stripe ;
   antennae in ♂ not or very little longer than in the ♀ ; maxillary palpi
   4 6 jointed . . . . .21.
Clypeus in ♂ more or less yellow (rarely black), in ♀ black ; antennae *in ♂
   very much longer than in the ♀* or deformed ; maxillary palpi 3, 4, 5 6
   jointed . . . . .18.
Clypeus in both sexes yellow or white; *antennae short and much alike in both
   sexes ;* recurrent nervures received respectively by the second and
   third cubital cells a little before the first and second cubiti.
Marginal cell at apex broadly obliquely truncate; dorsal abdominal segments broadly, sharply depressed at apex, the depressed part differently sculptured from the basal portion ; claws in ♀ simple, in ♂
   cleft ; eyes in ♂ convergent above ; mandibles bidentate ; flagellum
   subclavate . . . .**Meliturga** Latr.
Marginal cell at apex not truncate, narrowly rounded, or somewhat pointed ;
   dorsal abdominal segments at apex not depressed, uniformly sculptured ;
   claws cleft in both sexes ; eyes in ♂ not convergent above; mandibles
   simple ; flagellum filiform. . . . .**Meliturgopsis** Ashm. n. g.
18. First and second cubital cells, along the cubitus, equal or nearly, considerably
   longer than wide, but shorter than the third ; first recurrent nervure

received by the second cubital cell beyond the middle ; mandibles at apex bluntly rounded, edentate ; maxillary palpi 3-jointed.

Epimelissodes Ashm. n. g. (Type *M. atripes* Cr.).

First and third cubital cells, along the cubitus, subequal, the second smaller or shorter than either the first or third, the third the longest.

First recurrent nervure received by the second cubital cell at or near its middle or beyond the middle, the second recurrent received by the third cubital cell near its apex or almost interstitial with it ; third transverse cubitus not strongly angulately bent, the knee formed by the curvature of the vein towards the radius rounded.

Maxillary palpi 6-jointed ; submedian and median cells equal ; abdomen in ♀ black, not fasciate · **Synhalonia** Patton.

Maxillary palpi 5-jointed ; submedian cell a little shorter than the median.
Xenogloss Smith.

Maxillary palpi 4-jointed ; abdomen fasciate or subfasciate (rarely without) · **Melissodes** Latr.

First recurrent nervure received by the second cubital cell *much beyond its middle* or its apex, the second recurrent received by the third cubital cell just in front of the third transverse cubitus (only interstitial) ; third transverse cubitus strongly angulated or bent, its upper half or more bent inwardly towards the radius, so that the third cubital cell along the radius is only one-half as long (or even less) than along the cubitus ; first discoidal cell not longer than the marginal cell, most frequently somewhat shorter.

Maxillary palpi 6-jointed ; abdomen in ⚥ not fasciate, in ♀ fasciate ; submedian cell a little shorter than the median.

Eusynhalonia Ashm. n. g. (Type *S. Edwardsii* Cr.).

Maxillary palpi 5-jointed, the last two joints united, scarcely longer than the third, first and second joints subequal, much longer than the third ; submedian cell very distinctly shorter than the median ; second cubital cell quadrangular, longer than wide, but shorter than the first or third ; abdomen fasciate, or in ♀ clothed with a fine, short, whitish or pruinose pubescence.

Xenoglossodes Ashm. n. g. (Type *X. albata* Cr.).

Maxillary palpi 4-jointed : scopa on hind legs in ♀ long and densely plumose ; antennæ in ⚥ longer than the thorax, the clypeus yellow (rarely black).

Maxillary palpi with all the joints slender, the second and third elongate, the first and fourth very short ; abdomen fasciate. **Melissodes** Latr.

Maxillary palpi with the first joint enlarged, thickened, nearly as long as 2-4 united, the following slenderer, the second about as long as 3-4 united · **Ecplectica** Holmberg.

19. First and third cubital cells, along the cubitus, about of an equal length or nearly, the second much smaller, wider (higher) than long ; transverse median nervure interstitial or nearly, at most uniting with the median vein just in front of the basal nervure ; maxillary palpi 6-jointed · · 21.

20. First and second cubital cells subequal, the second a little the longer ; basal depression on first abdominal segment bounded superiorly by a distinct transverse carina ; hind tibiæ in ♀ with a dense scopa ; antennæ in ⚥ very long, much longer than in ♀, the clypeus yellow, maxillary palpi 5-jointed · · · · · · · · · · · · · · · **Eucera** Scopoli = *Anthophorula* Ckll.

21. Stigma more or less developed, at least twice as long as wide.

Abdomen smooth, polished, or microscopically reticulated, with hair bands, the venter with a long, sparse pubescence; hind legs in ♀ with rather long, sparse, black scopa, but conspicuously plumose; tongue much elongate, extending to apex of the first abdominal segment; maxillary palpi with joints 1-2 of an equal length; first joint of labial palpi scarcely half the length of the second ciliated; basal joint of hind tarsi in ☿ long, slightly curved, but not produced into a process beyond the insertion of the second joint.....**Entechnia** Patton.

Abdomen rarely polished, densely pubescent and usually fasciate, the venter with a long, dense pubescence; hind legs in ♀ with a long, dense, whitish, griseous or yellowish scopa; tongue not greatly lengthened; joints 2 and 3 of maxillary palpi almost twice as long as the first; first joint of labial palpi longer than the second; basal joint of hind tarsi in ☿ very long, strongly curved and produced into a long process beyond the insertion of the second joint.

Ancyloscelis Latr. = *Diadasia* Patton.

22. Second submarginal cell, along the cubitus, longer than wide............23.

Second submarginal cell, along the cubitus, wider (higher) than long and scarcely half the length of the first.

Transverse median nervure not interstitial, but joining the median vein a little beyond the origin of the basal nervure; first recurrent nervure interstitial or nearly with the second transverse cubitus; second recurrent nervure received by the third cubital cell at its apex or just before the third transverse cubitus; hind tibiae with a long, dense scopa; maxillary palpi 6-jointed; labial palpi 4-jointed.

Exomalopsis Spinola.

23. First recurrent nervure received by the second cubital cell a little *before* its middle, the second recurrent received by the third near its apex; maxillary palpi 6-jointed; labial palpi? 2-jointed.

Tetrapædia Klug.

First recurrent nervure received by the second cubital cell distinctly beyond its middle, or near its apex, nearly interstitial with the second transverse cubitus; second recurrent nervure almost interstitial with the third transverse cubitus; basal depression of first segment bounded by a transverse carina superiorly; maxillary palpi 6-jointed, joints 1-4 subequal in length, 2-6 shorter; clypeus in male yellow.

Diadasiella Ashm. n. g. (Type *D. coquilletti* Ashm.).

Family VI NOMADIDÆ.

(The Cuckoo Bees).

To this family belong the cuckoo or parasitic bees, with three submarginal cells in the front wings. Most of the species are bright colored or metallic-blue or green, with the abdomen most frequently marked with white pubescent maculæ, banded, or ornate with yellow or white.

All, without exception, live parasitically in the nests of other bees, and have undoubtedly originated from other bees, through different lines of descent. It is evident, however, that most of them are descendents from various Anthophorid bees, since they agree more nearly with these bees in venation and the characters of the mouth parts than with any of the others.

They are easily distinguished from the Anthophoridæ, however, by color, by the ♀ having no polleniferous scopa, or at most with only a very short, sparse pubescence, and by their bodies being most frequently bare or nearly, the pubescence, if any, being short and sparse. Very rarely are they densely pubescent on the head and thorax, as in the Anthophoridæ. The species are metallic, or rufous and black, or rufous and yellow, the abdomen being most frequently ornate with yellow or white maculæ or bands.

Some of them also resemble quite closely some of the Stelididæ, another family of parasitic bees; but the latter have only two submarginal or cubital cells in the front wings, while the labrum is, as a rule, not so well developed nor so prominent and always strongly inflexed. I believe also that the Stelididæ had quite a different line of descent, or from the Megachilidæ, their characters agreeing more nearly with this family than with any other.

The numerous genera, now placed in the Nomadidæ, may be distinguished by the use of the following table:

Table of Genera.

Marginal cell at apex not separated from the costa....................17.
Marginal cell at apex rounded, always separated from the costa or truncate.
    Cubital cells, along the cubitus, nearly of an equal length, the first as a rule somewhat the longest cell....................14.
    Cubital cells, along the cubitus, of an unequal length, one or another most frequently longer or much smaller, the first usually much the longest, or at least somewhat the longest (very rarely with the third the longest).
        Third cubital cell, along the cubitus, longer than either the first or second ....13.
        Third cubital cell, along the cubitus, equal to the first or distinctly shorter ....2.
2. First cubital cell, along the cubitus, much longer than the third, sometimes as long as the second and third united............................3.
  First cubital cell, along the cubitus, equal or nearly equal to the third, never very much longer.
    Second cubital cell either petiolate or very much narrowed above; marginal cell elliptical, not or only a little longer than the first cubital cell, or less than half the length of the first discoidal cell; scutellum subbilobed; axillæ rounded behind; abdomen short, subglobose as viewed from above; labial palpi 3-jointed, the first joint stout, longer than 2-3 united; claws with a tooth or cleft.............**Zacosmia** Ashm.

Second cubital cell quadrate or trapezoidal, as long as the first or third, conical, banded or marked with white pubescence; maxillary and labial
palpi 4-jointed.............................**Leiopodus** Smith.

3. Second cubital cell, along the cubitus, shorter than either the first or third, or
the second and third are equal or nearly.

Third cubital cell, *along the radius*, much shorter than along the cubitus, being
narrowed above...16.

Third cubital cell, *along the radius*, as long as along the cubitus, not narrower
above than below...4.

4. First cubital cell, along the cubitus, much shorter than the second and third..9.

First cubital cell, along the cubitus, as long as the second and third united or
nearly.

Third transverse cubitus angularly broken, not strongly curved outwardly..8.

Third transverse cubitus not angularly broken, strongly curved outwardly..5.

5. Submedian cell somewhat longer than the median, but the transverse median
nervure is straight, oblique or only slightly bent, but not angulated..7.

Submedian cell much longer than the median, the transverse median nervure
strongly angulated or > shaped.................. ....6.

6. Body black, the thorax densely pubescent, with griseous or pale fulvous
hairs above.

Abdomen black, immaculate; scutellum broad, the axillæ produced posteriorly into long, acute teeth or spines, the postscutellum unarmed.
                                    Bombomelecta Patton.

Abdomen black, shining, but the dorsum with maculæ of white pubescence
laterally; scutellum somewhat convex, scarcely bilobed, the postscutellum armed with a spine or tooth, the axillæ normal, unarmed;
maxillary palpi 5-jointed; labial palpi 4-jointed.....**Melecta** Latr.

Body rufous or rufous and black, bare or nearly; scutellum bilobed, the axillæ
triangularly acute at apex, but neither long nor extending beyond the
hind margin of the middle lobe.

Pyrrhomelecta Ashm. n. g. (Type *Epeolus glabrata* Cr.).

Body black, the thorax almost bare above, the abdomen dorsally banded with
white or with white lateral maculæ; axillæ triangular; scutellum
proper middle lobe) very large, flat, quadrangular and extending
over the metathorax, the hind margin sinuate; labial palpi 2-jointed;
maxillary palpi 5-jointed......:.......................**Crocisa** Latr.

7. Third cubital cell not narrowed above, the marginal cell not longer than half
the length of the first discoidal cell.

Scutellum bilobed, each lobe with a short conical tooth on its disk; axillæ
normal; postscutellum unarmed; labial palpi 5-jointed; maxillary
palpi 4-jointed, the last joint minute...**Pseudomelecta** Radoszk.

Scutellum terminating in two pointed tubercles; axillæ produced posteriorly
into acute incurved spines; postscutellum unarmed; labial palpi
4-jointed; maxillary palpi 2-jointed, the basal joint minute globose.
                                    Thalestria Smith.

Third cubital cell narrowed above, being much shorter along the radius than
along the cubitus, the marginal cell longer than half the length of the
first discoidal; labial palpi 2-jointed.

Metathoracic angles densely pubescent; transverse median nervure not angulate; scutellum bilobed.......................**Ericrocis** Cress.

Metathoracic angle bare or with a very short appressed pubescence; transverse median nervure angulate; scutellum convex, with a very slight median depression posteriorly, or at most sub-bilobed ; axillæ triangular, blunt behind.................................**Epeolus** Latr.

8. Mandibles simple ; abdomen with transverse bands.

Free portion of the marginal cell much less than that occupied by the cubitals.......................................**Dœringiella** Holmberg.

Free portion of the marginal cell much greater than that occupied by the cubitals ; second cubital cell subtriangular ; maxillary and labial palpi as in *Dœringiella*..............**Trophocleptria** Holmberg.

9. First recurrent nervure not interstitial with the second transverve cubitus, but joining the second cubital cell *before* or *beyond* the middle.........11.

Frst recurrent nervure interstitial with the second transverse cubitus.

First and second cubital cells, along the cubitus, of an equal length or nearly, the second sometimes a little the longer ; labial palpi 4-jointed....10.

First cubital cell, along the cubitus, longer than the second, the latter hardly so long as the third ; scutellum normal ; maxillary palpi 4-jointed.

Rhathymus Lepel. = *Lioguster* Perty.

10. Second cubital cell, *along the cubitus*, twice as long as the third, the latter wider than long; scutellum bituberculate ; maxillary palpi 1-jointed, with an annulus at base............**Eurytis** Smith = *Hopliphora* Lepel.

Second cubital cell, *along the cubitus*, rarely much longer than the third ; scutellum bilobed ; maxillary palpi 3-jointed..........**Melissa** Smith.

11. First recurrent nervure joining the second cubital cell *at* or *before* the middle..12.

First recurrent nervure joining the second cubital cell *beyond* the middle.

Scutellum simple, rounded posteriorly ; mandibles bidentate ; maxillary palpi 4-jointed ; labial palpi 2-jointed.....**Melectoides** Taschenb.

Scutellum bituberculate.

Second cubital cell smaller than the third, both a little narrowed above, the second along the radius scarcely two-thirds the length of the third ; tibial spurs simple..........................**Mesocheira** Lepel.

Second cubital cell longer than either the first or third, oblong-quadrate, the third petiolate ; hind tibial spurs simple ; maxillary palpi 6-jointed ; labial palpi 4-jointed....................**Mesonychium** Lepel.

Scutellum distinctly bidentate.

Second cubital cell almost quadrate, a little smaller than the third ; middle tibiæ with one spur, its apex bifid, the posterior tibial spurs serrate within.................................**Mesocheira** Lepel.

12. Marginal cell separated from the costa its entire length; second recurrent nervure interstitial with the second transverse cubitus ; maxillary palpi 6-jointed ; labial palpi 4-jointed..........**Florentina** D. T.

Marginal cell separated from the costa only towards the apex ; second recurrent nervure not interstitial.

Labrum longer than wide.

Maxillary palpi wanting ; labial palpi appearing as a single long bristle ; mandibles bidentate..**Exærete** Hoffmansegg = *Chrysantheda* Perty.

Maxillary palpi 2-jointed ; second cubital cell trapezoidal ; scutellum normal, subconvex......................**Epeicharis** Radoszk.

Maxillary palpi 3-jointed ; second cubital cell quadrate ; scutellum elevated ....**Epicharoides** Radosz.

Labrum wider than long.

Maxillary palpi 6-jointed ; second cubital cell shorter than the third ; scutellum not elevated ; mandibles bidentate..... **Epeoloides** Girard.

13 Second and third cubital cells each receiving a recurrent nervure........16.

Third cubital cell receiving both recurrent nervures; middle tibial spur dilated and serrated ♀ ....**Acanthopus** Klug.

14. Third cubital cell receiving both recurrent nervures; antennae very long, longer than the body ♂.

                   Acanthopus Klug. = *Ctenioschelus* De Roman.

Second and third cubital cells each receiving a recurrent nervure.

Metallic bluish green species; clypeus mith a median depression, but not bounded by distinct carinae laterally ; the clypeus itself, however, laterally before its apex has a carina, which curves towards the base of the eyes; scutellum large, with a ridge or elevation at sides; maxillary palpi 2-jointed ; labial palpi 4-jointed.............**Aglæ** Lepel.

15. First and second cubital cells subequal, shorter than the third, the second trapezoidal ; non-metallic species.

Labrum normal, without a median depression, yellow; clypeus and abdomen yellow, the latter banded with black ; antennae in ♂ deformed, the scape very stout, the flagellum subcylindrical, tapering towards apex, the last joint produced into a curved spine, as in *Meidamea* Cr.; hind femora much swollen.

             Cænonomada Ashm. n. g. (Type *C. Brueri* Ashm.).

16. Second and third cubital cell, along the cubitus, about equal, shorter than the first ; second cubital cell quadrate.

Abdomen normal, neither lengthened nor narrowed ; hind femora, with a tooth beneath ; maxillary palpi 6-jointed ; labial palpi 4-jointed.

                       Lipotriches Gerst.

Abdomen much lengthened and narrowed ; ♀ with the terminal ventral segment much produced, forming an elongate receptable for the base of the sting, which is greatly exserted ; hind femora simple.

                       Osiris Smith.

17. Marginal cell not quite as long as the three cubital cells united, narrowly rounded at apex, and with a slight appendage ; second cubital cell quadrate, half the length of the first ; first recurrent nervure joining the second cubital cell near its apex, the second recurrent joining the third beyond its middle ; scutellum large, transverse quadrate greatly produced and extending over the metathorax, similar to *Crocisa*, depressed above, with its apical margin medially triangularly emarginate ; abdomen as in *Crocisa*, with oblong, pearly white spots at the apical angles of the dorsal segments; tongue long, densely pilose (Africa).....**Crocisaspidia** Ashm. n. g. Type *C. chandleri* Ashm.).

Marginal cell as long as the three cubital cells united, well rounded at apex ; third cubital cell oblong-quadrate, almost as long as the first and second united, the second cubital the smallest cell—scarcely half as long as the third and narrowed above; first recurrent nervure joining the second cubital cell a little beyond the middle, the second recurrent joining the third at its middle ; scutellum large, quadrate, with a tubercle on each side at base; abdomen narrow, acutely conical, the last ventral segment produced much as in *Osiris*.

                       Cœlioxoides Cress.

18. Submedian cell distinctly longer than the median; maxillary palpi 6-jointed.
                                           Nomada Scopoli.
Submedian cell not longer than the median; maxillary palpi 5-jointed.
                                    Brachynomada Holmberg.

Family VII. CERATINIDÆ.

(The Small Carpenter Bees).

The bees placed in this family I have called "The Small Carpenter Bees," on account of their habits being similar to the large carpenter bees, or the family Xylocopidæ. Indeed, their relationship to this family is extremely close, although I consider them just as closely allied to certain Osmiines in the family Megachilidæ.

The European authority, Dr. Schmiedeknecht, has placed them in the family Xylocopidæ and this arrangement has been followed by Dr. Friese and others.

They are mostly small metallic blue, blue-black (rarely black), blue-green, or bright green bees, almost entirely devoid of pubescence and *without* a distinct polleniferous scopa on hind legs and venter, and in their very much smaller size and general appearance are so totally different from the large carpenter bees that I cannot believe they are at all related. I have therefore not hesitated, since the characters lacking in the species are of great taxonomic importance, to separate them as a distinct family.

Two genera can be distinguished, as follows:

Table of Genera

Third cubital cell, along the cubitus, fully as long as the first, but so much narrowed above, along the radius, that its length is reduced one-half; second cubital cell the shortest, narrowed above; second and third cubital cells each receiving a recurrent nervure beyond the middle; antennæ short, subclavate; head seen from in front a little longer than wide; maxillary palpi 4-6-jointed; labial palpi 4-jointed.
  Males; mandibles bidentate.................................. ....2.
  Females; mandibles tridentate.
    Maxillary palpi 4-jointed·
                Zadontomerus Ashm. n. g. (Type *C. tejonensis* Cr.)·
    Maxillary palpi 6-jointed.......................**Ceratina** Latreille.
2. Hind femora produced into a large triangular tooth beneath; genital ventral plate trapezoidal, not wider than long; maxillary palpi 4-jointed.
                                    Zadontomerus Ashm.
  Hind femora normal; genital ventral plate semicircular, twice wider than long; maxillary palpi 6-jointed..............**Ceratina** Latreille.

Family VIII. XYLOCOPID.E.

(The Large Carpenter Bees).

The bees belonging in this family are for the most part very large, robust bees, having the head and thorax, especially laterally, clothed with a rather dense pubescence, the abdomen convex above, with a ventral scopa, while the hind tibiæ and tarsi in the females are furnished with a dense polleniferous scopa.

These bees closely resemble the bumble bees, and some of the largest bees, if not the largest bees known, belong to it.

I have included with them two genera of uncertain position, namely, *Oxæa* and *Lestes*, which probably should be considered as a subfamily Oxæinæ. They are placed here on account of possessing the long, narrow, marginal cell, which is characteristic of the family.

Dr. Henry Freise has quite recently placed *Oxæa* with certain Colletidæ, *Megacilissa*, *Caupolocana*, etc., but I cannot believe this to be its true position, since the mouth parts are totally different from these bees. Its affinities, it seems to me, are much closer to the Anthophoridæ, where some authorities have already placed it. I believe, however, its true position can only be settled definitely when its habits are known.

The genera which I have recognized in this family may be tabulated as follows:

Table of Genera.

Front wings with three cubital cells...2.
Front wings with two cubital cells.
    Thorax in ♀ with either a blue or bluish gray pubescence; maxillary palpi 5-jointed, the first joint short, stout, the second much the longest joint; labrum in ♀ trilobed; ♂ with the eyes somewhat convergent above, the tibiæ long and rather slender, the hind tibiæ at apex produced into a strong blunt process beneath (Java).
            Cyancoderes Ashm. n. g. (Type *C. Fairchildi* Ashm., also *X. cærulea* Fabr.).
2. Third cubital cell, along the cubitus, only slightly longer than the second, but along the radius a little shorter; second cubital cell oblong-quadrate, never triangular; legs clothed with moderately long, sparse hair....5.
    Third cubital cell much longer than either the first or second.
    Second cubital cell half the length of the first, quadrate or nearly; first recurrent nervure interstitial or nearly with the second transverse cubitus...4.
    Second cubital cell triangular; first recurrent nervure interstitial or nearly with the second transverse cubitus, sometimes entering the second cubital cell just in front of this nervure; second recurrent nervure received by the third cubital cell a little beyond the middle; eyes densely pubescent.

Scutellum posteriorly truncate, the apical margin acute and projecting over the metanotum ; basal abdominal segment broadly, deeply concave at base, the concavity superiorly acutely margined ; labial palpi 5-jointed, the three basal joints elongate, the last two minute; eyes extending to base of mandibles, in male more or less strongly convergent above.................... ....3.

Scutellum rounded off posteriorly, not projecting over the metanotum ; basal abdominal segments only slightly concave at base, the impression superiorly rounded, not acutely margined ; labial palpi 4-jointed ; eyes not quite extending to base of mandibles, in ♂ not strongly convergent above ; intermediate legs normal.

　　　　　　Xylocopa Latr. = *Shoruherria* Lep. (Type *X. violacea* Latr.).

3. Labrum in ♀ trilobed ; scape of antennæ cylindrical, not flattened.

Second cubital cell, along the cubitus, much longer than the first ; mandibles tridentate ; ♂ with the eyes not strongly convergent above, the intermediate legs deformed, their tarsi compressed, with a ,long. lateral fringe of hairs........**Mesotrichia** Westw. (Type *M. torrida* W.).

Second cubital cell, along the cubitus, shorter than the first ; mandibles bidentate ; ♂ with the eyes not or only slightly convergent, the intermediate legs normal, but both the intermediate and the anterior tarsi fringed with very long hairs........**Koptorthosoma** Gribodo.

Labrum in ♀ unilobed, or with a median carina on ridge ; scape of antennæ flattened ; second cubital cell, along the cubitus, about equal to the first ; ♂ with the eyes strongly convergent above, almost holoptic ; front coxæ armed with long spines beneath ; front tarsi broadly dilated.

　　　　　　Platynopoda Westw.=*Audineta* Lepel. (Type *X. latipes* Fabr.).

4. Maxillary palpi 4-jointed, gradually tapering from the base to apex, the basal joint stout, about half the length of the second, the third shorter than the basal joint, the apical joint slender, minute ; labial palpi 4-jointed, shorter than the labium...........................**Lestis** Lepel.

5. Maxillary palpi wanting ; first joint of flagellum, in both sexes, elongate, nearly as long as the scape ; eyes in ♂ strongly convergent and almost meeting above.......................... ....**Oxæa** Klug.

Family IX. MEGACHILIDÆ.

(The Mason, Leaf-cutting and Potter Bees).

The bees placed in this family are exceedingly common, and are found widely distributed into all regions of the globe. In the number of genera and species it will probably exceed all of the other bee families, except possibly the Anthophoridæ.

It seems to be dividable into three natural groups, which I have designated as subfamilies. These may be distinguished by the following simple characters :

Table of Subfamilies

Abdomen in ♀ always with a ventral scopa.

　　Abdomen above convex ; terminal tarsal joint always with a distinct pulvillus between the claws... ....Subfamily I. OSMIINÆ.

Abdomen above less distinctly convex, somewhat flat, depressed, rarely subconvex; terminal tarsal joint without a pulvillus between the claws.

Second cubital cell receiving both recurrent nervures; stigma poorly developed, narrowed, but about twice as long as wide; submedian and median cells most frequently equal, rarely with the submedian the longer; abdomen never banded or maculate with white or yellow, at most with hair bands............... Subfamily II. MEGACHILINÆ.

Second cubital cell receiving only one recurrent nervure—the first, the second recurrent joining the radius a little *beyond* the second transverse cubitus, or at most interstitial with it; stigma scarcely developed, at most not or scarcely longer than wide; submedian cell most frequently a little longer than the median, sometimes equal with it; abdomen above bare, always banded or maculate with yellow-white or rufous, never fasciate with hair bands.

Subfamily III. ANTHIDIINÆ.

Subfamily I. OSMIINÆ.

(The Mason Bees).

The bees placed in this subfamily are readily separated or distinguished from the next two families, into which I have divided the family, *by always having a distinct pulvillus between the claws*, a character not possessed by the other two subfamilies. They have, too, quite a distinct habitus of their own, scarcely definable, but readily recognizable by the experienced eye, their heads being slightly different in shape, with usually broader temples, the abdomen more convex above and rarely with distinct white hair bands as in the Megachilinæ. Their color, as a rule, is different, being more or less metallic, dense black, blue-black or blue, and through blue-green to a bright metallic green and cupreous.

The species in only a few genera resemble certain Megachilinæ, i. e., *Trypetes, Heriades, Chilostoma, Ashmeadiella*, etc., and these must be examined with care to distinguish them from some Megachilines.

The habits of the species in this group, so far as they are known, also support their separation as a distinct group from the others. Their nests are made in old posts, trunks and limbs of old and decaying trees, or in the interstices of stone walls, etc., the partitions between their cells being filled with clay and sand or other material, the cells themselves being thickly covered with sand externally.

The genera recognized are as follows:

Table of Genera.

Both recurrent nervures received by the second cubital cell, or the second recurrent is interstitial with the second transverse cubitus.

Claws in ♀ simple, neither cleft or without a subapical tooth, in ⚥ cleft or simple; face in ⚥ never marked with yellow or white............2.

Claws in ♀ cleft or with a distinct subapical tooth; in ⚥ cleft; face in ⚥ anteriorly marked with yellow or white.

Face in ⚥ yellow or white; pygidium broadly emarginate at apex, the hypopygium trapezoidal broadly emarginate: maxillary palpi 4-jointed; body above clothed with fulvous hairs; the abdomen short.

Trachusa Jarine = *Diphysis* Lepel.

Face in ⚥ with the clypeus anteriorly alone yellow; pygidium triangular entire, the hypopygium normal; body clothed with whitish or cinereous hairs, tinged with ochraceous on vertex and on thorax above, the abdomen with distinct white fasciae.

Zacesta Ashm. n. g. (Type *Z. rufipes* Ashm.).

2. Maxillary palpi 4- or 5-jointed...4.
Maxillary palpi 3-jointed.

Abdomen with the basal impression on the first abdominal segment not bounded superiorly by a distinct transverse elevated line or carina...3.

Abdomen with the basal impression on the first segment bounded superiorly by a distinct transverse elevated line or carina; mandibles tridentate : claws simple : abdomen most frequently fasciate with white.

Trypetes Schenck.

3. Mandibles at apex tridentate; clypeus simple; triangular area of the metathorax smooth, shining; ⚥ antennæ not crenulate beneath.

♀ abdomen elongate, longer than the head and thorax united; transverse median nervure in front wings quite interstitial; pygidium in ⚥ short, impressed above, but broadly, squarely truncate posteriorly, with acute angles; second ventral segment with a prominent tooth or ridge.

Heriades Spinola.

♀ abdomen broader and shorter, not longer than the head and thorax united, usually somewhat shorter; transverse median nervure in front wings not quite interstitial with the basal nervure, uniting with the median vein a little before the origin of the basal nervure: pygidium in ⚥ 4-dentate; second ventral segment normal.

Ashmeadiella Ckll.

Mandibles at apex bidentate: clypeus in ♀ with a lamina, tubercle or ridge anteriorly; triangular area of metathorax opaque, crenulate basally : ⚥ antennæ crenulate beneath, the pygidium deeply emarginate, bidentate or forked..**Chelostoma** Latr. = *Gyrodroma* Klug (Thoms.) = *Chelynia* Prov.

4 Maxillary palpi 5-jointed; flagellum in both sexes compressed : abdomen in ⚥ elongate, usually with the segments narrowly fasciate, the impression at base of first segment not broad, poorly defined, usually represented by a longitudinal sulcus..5.

Maxillary palpi 4-jointed ; abdomen most frequently shorter, not elongate, the first dorsal segment with a broad, subconcave impression at base, rarely otherwise.

Antennæ similar in both sexes, never deformed in the ♂ . . . . .6.

Antennæ dissimilar in the sexes, in ♂ deformed, in ♀ filiform, simple;
scape always stout; first dorsal abdominal segment at base convex not
broadly impressed, at most with a longitudinal sulcus; stigma in front
wings well defined.　　　　　　　　'

Flagellum in ♂ thickened, the joints compressed, submoniliform, nearly
of an equal length, the terminal joint abruptly constricted into a slender curved spine; scape large, robust; pygidium triangular, the apical
angles of the sixth dorsal segment dentate; second ventral segment
with a median prominence or ridge. . . . .**Alcidamea** Cress.

Flagellum in ♂ with the joints of an unequal length, joint 2-5 compressed,
dilated, joint 6 suddenly narrowed towards apex, the apical joint simple; scape rather long and robust, slenderer towards base; abdomen
elongate; pygidium obtusely triangular, with a deep transverse impression on disk, the apical lateral angles of the sixth segment dentate; first ventral segment subtriangularly produced at apex.

　　　　　　　　　　　　　　　　　Andronicus Cress.

5. Body black or blue-black, the thorax with a griseous pubescence, sometimes
mixed with black hairs; mandibles at apex very broad, 4-dentate, or
at least trisinuate; ♀ scopa black; the dorsal segments, except the
first or second at sides, not fasciate; ♂ abdomen most frequently with
narrow fasciæ, sometimes interrupted medially, the pygidium large,
twice as broad as long, with a depression on the disk, the apical margin
subarcuate; lateral angles of sixth segment acute.

　　　　　　　　　　　　　　　　　Monumetha Cress.

6. Malar space wanting, the eyes extending to the base of mandibles; ♀ with
the posterior orbits normal, not produced below into a tubercle behind
base of mandibles; antennæ in ♂ shorter than the thorax or no
longer. . . . . .7.

Malar space distinct, as long or a little longer than the pedicel, the eyes not
quite extending to base of mandibles; ♀ with the posterior orbits
produced at base into an angle or tubercle, usually a little back of the
mandibles and between them and the eyes; clypeus in ♀ deeply
emarginate anteriorly, and often armed with a horn, tooth or tubercle
on each side; ♂ with the antennæ longer than the thorax, the sixth
dorsal abdominal segment laterally and apically entire, not sinuate or
emarginate, the pygidium semi-circular or obtuse-triangular, entire or
at least never deeply emarginate at apex, at most with a very slight
sinus. . . . . **Ceratosmia** Thoms.

7. Tibial spurs pale or rufous, never black or blue-black; abdomen most frequently not metallic, very rarely distinctly metallic; ventral scopa in
♀ white, rufous or fulvous, never black; ♂ with the sixth dorsal
abdominal segment laterally sinuate, or emarginate and frequently
with a lateral tooth, the pygidium at apex emarginate, bi- or tri-den-
tate, rarely entire. . . . . .9.

Tibial spurs black or blue-black; body metallic, or at least the abdomen is
metallic, black, blue-black or æneous black.

Ventral scopa in ♀ rufous or black only at apex. . . . . .8.

Ventral scopa in ♀ white. . . . . .8½.

Ventral scopa in ♀ black.

Mandibles long, rostriform or subrostriform, at apex forcipate; maxillary palpi rather long; metathorax with its basal area opaque or nearly : labrum with two fascicles of black or rufous hairs (one on each side), which project from under the clypeus; ♂ with the antennæ short, the sixth dorsal abdominal segment laterally entire, not sinuate, and with no sinus or incision at apex, the pygidium trapezoidal about twice as wide as long, the apex with only a slight emargination, never deeply emarginate...............................**Melanosmia** Schmiedek.

Mandibles not long rostriform, very broad, 4-dentate, and usually with some rufous hairs above towards apex ; maxillary palpi short ; clypeus with a median production anteriorly ; ♂ with the antennæ shorter than the thorax, the sixth dorsal abdominal segment laterally sometimes sinuate, but not toothed, the pygidium deeply emarginate at apex....................**Osmia** Panzer = *Chalcosmia* Schmiedek.

8. Mandibles obtuse or 4-dentate; body black, densely pubescent; ♀ with the antennæ short, the flagellum compressed or subcompressed.

♂ with the antennæ very short, compressed and much dilated towards the base, the third ventral segment armed with a long spine, the pygidium quadrate...............................**Arctosmia** Schmiedek.

♂ with the pygidium entire; mandibles 4-dentate, forcipate ; body blue-black or æneous black, with fulvous pubescence.

Acerotosmia Schmiedek.

8½. Mandibles in both sexes 3-dentate, the apical tooth acute ; clypeus slightly produced anteriorly and squarely truncate; maxillary palpi short : body metallic, or at least submetallic; anterior tibial hook distinct, acute ; ♂ with the sixth dorsal abdominal segment laterally most frequently sinuate, but not toothed, with a slight or distinct median emargination apically, the pygidium deeply semi-circularly emarginate, bidentate.

Nothosmia Ashm. n. g. (Type *O. distincta* Cr.).

9. Mandibles differently shaped, not subrostrate ; maxillary palpi short, the last joint minute, subobsolete...............................10.

Mandibles subrostrate, forcipate ; maxillary palpi rather long ; body black. subæneous, and more or less densely pubescent, the abdomen fasciate : metathoracic basal area opaque ; scopa in ♀ rufous ; ♂ with the sixth dorsal abdominal segment at apex emarginate and laterally deeply sinuate and dentate, the pygidium bidentate.

Amblys Klug = *Helicosmia* Thoms. (Type *O. bicornis* L.).

10. Abdomen black or metallic, never rufous; ventral scopa in ♀ white, or more rarely rufous ....11.

Abdomen rufous or rufous and black, punctate; ventral scopa in ♀ white or fulvous; pygidium in ♂ bilobed.

Pseudosmia Radoszk. = *Erythrosmia* Schmiedek.

11. Axillæ posteriorly normal, not produced into a spine ; occipital margin not acutely rimmed ; ventral scopa in ♀ white or rufous...............12.

Axillæ produced into acute teeth posteriorly, which extend beyond the post-scutellum ; anterior tibial hooks wanting ; occipital margin superiorly acutely margined : ventral scopa in ♀ rufous ; ♂ with the sixth dorsal abdominal segment at apex serrated, the first or second ventral segment sometimes spined...............................**Hoplosmia** Thoms.

12. Hind tibial spur in ♀ normal; abdomen subdepressed, thinly fasciate with
    white hairs; ♂ with the second ventral segment simple, not produced
    into a spine..13.

Hind tibial spur in ♀ broad, sulcate beneath; body black, the abdomen sub-
    glabrous, distinctly punctate and usually subfasciate, the ventral
    scopa rufous; clypeus at apex simple, the disk sometimes with a
    smooth impression; mandibles 3-dentate; anterior tibial hook wanting
    or indistinct; ♂ with the second ventral segment armed with a long
    spine or tooth.

    ♂ pygidium elongate-quadrate, the apex slightly rounded, not emarginate
       or forked....... ....**Acanthosmia** Thoms.
    ♂ pygidium trapezoidal and deeply emarginate at apex, therefore bidentate,
       the sixth dorsal segment laterally slightly sinuate, but entire at apex,
       although on its disk apically is a slight median furrow or sulcus.
       Acanthosmioides Ashm. n. g. (Type *O. odontogaster* Ckll.).

13. Body black, not at all metallic; pygidium in ♂ not tridentate............14.
    Body metallic; pygidium in ♂ tridentate.
    Clypeus in ♀ subemarginate or truncate, in ♂ tridentate; thorax above
       clothed with dense fulvous hairs; abdomen fasciate, the ventral scopa
       rufous, rarely whitish; mandibles in ♀ 3-dentate, in ♂ bi-dentate;
       pygidium in ♂ deeply bi-emarginate, forming 3 teeth, the median
       tooth acute, the lateral teeth rounded at apex, the sixth dorsal seg-
       ment laterally sometimes dentate.....**Tridentosmia** Schmiedek.

14. Clypeus at apex serrate-crenulate; anterior tibial hook distinct; front wings
       with the transverse median nervure uniting with the median vein a
       little *before* the origin of the basal nervure or interstitial; mandibles
       in ♀ 3-dentate, in ♂ 2-dentate; ♂ with the pygidium semi-circular,
       entire, the sixth dorsal abdominal segment laterally strongly emargin-
       ate, forming a tooth.
       Hoplitis Klug = *Ctenosmia* Thoms. (Type *O. adunca* Panz.)
    Clypeus anteriorly rounded, unarmed; anterior tibial hook generally wanting
       or poorly developed, obtuse; front wings with the transverse median
       nervure uniting with the median vein a little *beyond* the origin of the
       basal nervure; mandibles in both sexes 3-dentate; ♂ with the pygi-
       dium emarginate at apex or forked, the sixth dorsal abdominal seg-
       ment laterally sinuate, dentate.
       Anthocopa Latreille = *Furcosmia* Schmiedek.
    Clypeus at apex unarmed; body black, with fulvous hairs, the abdomen
       fasciate, the scopa rufous; mandibles 4-dentate; ♂ with the pygidium
       with a deep impression or fovea on its disk, the sixth dorsal abdominal
       segment laterally dentate, the middle femora armed with a tooth
       beneath................................**Megalosmia** Schmiedek.

Subfamily II. MEGACHILINÆ.
(The Leaf-cutting Bees).

The bees placed in this subfamily closely resemble those in the
former, and it requires long practice and an experienced eye to
separate some of them from each other, although the head, as a rule,

is more transverse, the temples not so broad, the stigma in front wings poorly developed, but narrower and longer, while the abdomen above is more depressed, most frequently distinctly fasciate or with white hair bands, the basal segment sharply truncate or broadly concave at base, so as to fit close to the metathorax when elevated. The absence of pulvilli between the claws is, however, the only reliable character that will separate them.

To this group or subfamily belong the genuine leaf-cutting bees, so called from the habit the female has of cutting small, almost circular pieces out of the tender leaves of various trees and plants, wherewith to line its cells. The cells themselves are cylindrical, tubular, or, in outline, not unlike a small open-mouthed vial, composed of numerous layers of pieces of leaves, wrapped into shape, layer upon layer, as a cigarmaker wraps his cigar; these cells are arranged in rows, end to end, one upon each other, in burrows or tunnels made in the ground or in decaying wood.

The genera are not numerous and may be recognized by the use of the following table :

Table of Genera.

Second cubital cell receiving both recurrent nervures.
  Mandibles strong, broad, dilated at apex and 3-, 4- or 5-dentate............2.
  Mandibles narrower, simple, bi- or tri-dentate, never broad or dilated at apex.
    Mandibles simple or bidentate at apex : ♀ with a prominent lamina beneath the insertion of antennae ; pygidium in ♂ simple or trilobed ; maxillary palpi 6-jointed ; labial palpi 4-jointed.
    ♀ with the inner spur of hind tibiæ lunulate and finely serrated within : apical abdominal segments in ♂ trilobed ; mandibles simple.
            Ctenoplecta Smith.
    ♀ with the inner spurs of hind tibiæ normal ; apical abdominal segment in ♂ unarmed, the antennae long, the apical two joints compressed, spatulate ; mandibles in ♂ simple, in ♀ bidentate.
            Steganomus Ritsema = *Cyathocera* Smith.
  Mandibles tridentate or subtridentate ; ♀ with a prominent lamina or ridge beneath the insertion of the antennae ; pygidium in ♂ terminating in a tooth or strong spine, the antennae normal ; maxillary and labial palpi 4-jointed..................................**Liturgus** Latr.
2. First cubital cell not longer than the second, of an equal length or nearly ; maxillary palpi 4-jointed...............**Megachile** Latr. (sens lat).
  First cubital cell distinctly longer than the second..........................3.
3. Marginal cell appendiculate at apex, or at least with a stump of a vein......4.
  Marginal cell not appendiculate at apex.
    Labial palpi 4-jointed, the two basal joints elongate, 3 4-jointed, subclavate ; maxillary palpi 2-jointed, minute ; antennae long, joints 9 11 very long and slender, 12-13 abruptly clavate......**Thaumatosoma** Smith.

4. Abdomen above highly convex; mandibles with the outer tooth strong and very acute, usually with a pencil of long ferruginous hairs before tips; outer apical edge of tibiæ acute; claws long, acute, simple.

Chalicodoma Lepel.

Subfamily III. ANTHIDIINÆ.

(The Potter Bees).

The bees belonging here are always brightly colored or ornate, and are at once distinguished from those in the two preceding subfamilies by having the second recurrent nervure uniting with the radius *beyond* the second transverse cubitus or cubital cell, or at most *interstitial* with the second transverse cubitus. In no species that I have examined have I found a single specimen with the second cubital cell receiving both recurrent nervures.

All of the bees belonging here are also readily distinguished from the others by their color being either black and rufous or rufous and yellow, with the abdomen always banded or maculate with yellow, white or rufous.

The habits of the group, too, curiously enough, is also quite different from the others. The female, normally, constructs a globular cell, not unlike in appearance to that of an Eumenid, but much smaller, attached to the stem of a plant, and made of a waxy-like substance and down stripped from pubescent or woolly-leafed plants.

Only three genera have been recognized, distinguished as follows:

Table of Genera.

Second recurrent nervure joining the radius behind the second transverse cubitus. . . . . .2.
Second recurrent nervure interstitial with the second transverse cubitus.
     Maxillary palpi 2-jointed; labial palpi 4-jointed; abdomen more or less ornate with tufts or fasciæ of white pubescence; the seventh segment in ♂ with a slender median spine and with two longer, stouter lateral spines . . . . **Scrapis** Smith.
2. Maxillary palpi 2-jointed; labial palpi 1-jointed; abdomen red or ferruginous, not spotted or fasciate with yellow; ♂ abdomen at apex tridentate.

Euaspis Gerst.
     Maxillary palpi 1-jointed; labial palpi 4-jointed; abdomen most frequently black, rarely ferruginous, fasciate or maculate with yellow, rarely ornate with white; ♂ abdomen nearly always toothed or spined at apex . . . . **Anthidium** Fabr.

Family X. STELIDÆ.

(Parasitic Bees).

This family is composed of genuine parasitic bees, living—like the Nomadidæ—in the nests of other bees, species belonging to it having been bred from the cells of *Anthophora*, *Megachile*, *Anthidium*, *Osmia*, etc. All are without a ventral scopa, and also without scopa on the hind legs.

The species composing it come nearest to the Megachilidæ, and some of the genera, as has been suggested by Dr. Friese, are offshoots from some of these bees.

The parasitic habits, so noticeable in various groups of insects, primarily or originally, I think, must have been evolved or developed independently, through lack of sufficient food supply, until it became acquired and hereditary in the offspring,

The family is dividable into two subfamilies:

Table of Subfamilies.

Claws with pulvilli between......................Subfamily I. STELIDINÆ.
Claws without pulvilli between....................Subfamily II. CŒLIOXINÆ.

Subfamily I. STELIDINÆ.

The bees placed in this group, as with the Osmiinæ in the Megachilidæ, always have a distinct pulvillus between the claws, which readily separate them from those found in the Cœlioxinæ.

Dr. Henry Friese suggests that the group, or at least the genus *Stelis*, originated from *Anthidium*.

Five genera are now recognizable:

Table of Genera.

Second cubital cell receiving only one recurrent nervure, the second recurrent uniting with the cubitus behind the transverse cubitus..............2.
Second cubital cell receiving both recurrent nervures.
   Abdomen black or blue-black, usually with white transverse bands or maculate; mandibles tridentate; maxillary palpi 2-jointed; ♂ with the pygidium subemarginate, the hypopygium tridentate.
                                                    Melanostelis Ashm.
2. Abdomen black or rufous and black, immaculate, clothed with a scattered pubescence; scutellum rounded and produced behind over the base of the abdomen..... ....4.
   Abdomen black or rufous, and most frequently ornate with white or yellow spots; scutellum quite differently shaped, normal or with lateral teeth; maxillary palpi 1 or 2-jointed (rarely wanting); labial palpi 4-jointed.
   Scutellum with lateral teeth behind; head narrower than the thorax.....3.
   Scutellum without lateral teeth behind; head fully as wide as the thorax.
      Clypeus not lengthened, well rounded; maxillary palpi 1 or 2-jointed; abdomen semiglobose, the segments broadly banded with yellow or white, as in *Anthidium*, ♂ with the anal segment entire, rounded, ending in a strong thorn.... ....**Protostelis** Friese.

Clypeus lengthened and deeply emarginate; maxillary palpi 1-jointed; abdomen longer, cylindrical, with large white spots, in ♀ with the apical margin of the sixth segment toothed, in ♂ with the seventh armed with a tooth..................**Stelidomorpha** Moraw.

3. Clypeus rounded, not lengthened; maxillary palpi 2-jointed; abdomen rounded or oval, black, rarely with small, indistinct maculæ; ♂ with the seventh segment rounded...........................**Stelis** Panzer.

4. Apex of scutellum with a deep median depression; maxillary palpi 2-jointed; labial palpi 4-jointed; mandibles 3-dentate; apical abdominal segment in ♂ tridentate............................**Parevaspis** Ritsema.

Subfamily II. CŒLIOXINÆ.

The bees composing this group are very much commoner and much more numerous in genera and species than are the Stelidinæ. Their habitus also is quite different; as a rule, more robust, less ornate, black, or at most with a rufous abdomen, the segments being somewhat constricted at sutures, or, if not, the abdomen is acutely pointed at apex; claws always *without* pulvilli between.

Dr. Friese suggests that *Cœlioxys, Dioxys,* etc., originated from *Megachile.*

The genera are fairly numerous, separated as follows:

Table of Genera.

Marginal cell at apex more or less remote from the costa, or somewhat truncate..2.
Marginal cell at apex attaining the costa.

> Median and submedian cells about equal; mandibles simple; hind femora normal.........................**Hypochrotœnia** Holmberg.
> Median and submedian cells unequal; mandibles at apex bidentate; hind femora much swollen, their tibiæ dilated, the tarsi very long.
> **Œdiscelis** Philippi.

2. Marginal cell at apex acuminate or narrowly rounded; mandibles usually dentate.. 3.
Marginal cell at apex truncate or broadly rounded most frequently with a slight appendage; mandibles acute or at least not dentate; abdomen rufous or rufous and black, ornate, rarely entirely black..........5.

3. First cubital cell distinctly longer than the second.......................8.
First cubital cell not longer than the second, equal or distinctly shorter; third antennal joint normal...4.

4. First cubital cell equal to the second or only slightly shorter; second cubital cell receiving both recurrent nervures, the first near its basal angle, the second before its apex; submedian cell much longer than the median; abdomen short, elliptic or oval-lanceolate; labial palpi 4-jointed, the first more than twice longer than the second, obconical, the last hardly longer than the third, fusiform........**Pseudopeolus** Holmberg.
First cubital cell distinctly shorter than the second, the latter receiving both recurrent nervures, the first at its basal third, the second just before the second transverse cubitus; submedian and median cells of an equal length; abdomen oval....................**Biastes** Panzer.

5. Scutellum bilobed or strongly bituberculate ; postscutellum unarmed........6.
  Scutellum simple, not distinctly bilobed.
    Postscutellum armed with a small tooth.
      Antennae somewhat distant at base ; maxillary palpi 6-jointed.
        Ammobates Latr. = *Phileremus*.Latr.
      Antennae strongly approximate at base.
        Maxillary palpi 3-4-jointed.............**Ammobatoides** Schenck.
        Maxillary palpi 2-jointed...................**Melanempis** Sauss.
    Postscutellum unarmed,
      First cubital cell twice the length of the second ; median and submedian
      cells nearly equal ; first recurrent nervure interstitial ; claws with a
      median tooth..............................**Chilicola** Spinola.
6. Both recurrent nervures received by the second cubital cell.................7.
  First recurrent nervure received by the first cubital cell, or interstitial with
  the first transverse cubitus.
    Marginal cell very small and short, scarcely the length of the stigma.
      Front wings with two complete cubital cells and three discoidal cells.
        Neolarra Ashm.
      Front wings with only one complete cubital cell and two discoidal cells.
        Phileremulus Ckll.
    Marginal cell not small, but long, much longer than the stigma.
      Front wings with the normal number of cells, the first cubital cell longer
      than the second, the second narrowed above ; labial palpi 4-jointed,
      maxillary palpi 6-jointed.
        Neopasites Ashm. (Type *P. fulviventris* Cr.).
7. Marginal cell much longer than the stigma.
  First cubital cell much longer than the second.
    Labrum nearly three times as long as wide ; maxillary palpi wanting ; labial
    palpi 2-jointed ; antennae 12-jointed in both sexes..**Pasites** Jurine.
    Labrum triangular ; labial palpi (?) 5-jointed.
      Schmiedeknechtia Friese.
    Labrum quadrate, with a delicate, median carina anteriorly ; maxillary
    palpi 6-jointed ; third antennal joint elongate, nearly thrice as long as
    the fourth......................................**Phiarus** Gerst.
  First cubital cell somewhat shorter than the second ; labrum subquadrate ;
  maxillary palpi 4-jointed, the first joint elongate, slender, the last as
  long as 2-3 united.......**Homachthes** Gerst. = *Morgania* Smith.
8. Axillae produced posteriorly into acute teeth ; eyes hairy or glabrous, in the
  latter case the postscutellum armed with a tooth or spine.........11.
  Axillae normal, not acutely toothed ; eyes always bare ; both recurrent nervures
  received by the second cubital cell, or rarely with the first recurrent
  nervure interstitial or received by the first cubital cell just before the
  first transverse cubitus.
    First cubital cell twice as long as the second, the second recurrent nervure
    received at the middle of the second cubital cell..................10.
    First cubital cell not twice as long as the second.
      Median and submedian cells unequal.............................9.
      Median and submedian cells equal, the transverse median nervure inter-
      stitial with the basal nervure ; maxillary palpi 6-jointed ; labial palpi
      4-jointed.

Front wings with two recurrent nervures, both received by the second
cubital cell............................. ....**Allodape** Lepel.
Front wings with only one recurrent nervure, the second wanting, the
third discoidal cell absent.....................**Exoneura** Smith.
9. Second cubital cell receiving both recurrent nervures.

Abdomen black or rufous and black, almost glabrous; venter in ♀ naked,
the anal segment excised ; ♂ antennæ 12 13 jointed ; scutellum bituberculate; postscutellum unarmed ; maxillary palpi 4-jointed.
Melittoxena Morawitz Type *M. truncata* Nyl.) = ? *Nomadita* Mocs.
Abdomen red and black, opaque, closely and densely punctate, the dorsal
segments at apex banded with an appressed, whitish pubescence ; no
transverse furrow at base of segments; axillæ acute or toothed at
apex; postscutellum armed with a median tooth.
Hoplopasites Ashm.
Second cubital cell receiving only one recurrent nervure—the second, the first
recurrent being received by the first submarginal cell at its apex.

Abdomen red ; dorsal segments 2 5 and the ventral segments 2 4 with deep,
transverse furrows at base ; pygidium with a median carina ; scutellum
unarmed............................**Holcopasites** Ashm n. g.
10. Median and submedian cells equal or nearly.

Maxillary palpi 5-jointed, the joints slender, elongate ; labial palpi 4-jointed,
the first joint large, elongate.............**Cænoprosopis** Holmb.
11. Eyes glabrous; postscutellum armed with a tooth or spine; teeth of mandibles unequal, the apical tooth much the longer.

♀ with the last abdominal segment truncate, in ♂ with the penultimate
and ultimate ventral segments with a lateral apical tooth.
Dioxys Lepel.
♀ with the last abdominal segment acuminate, in ♂ with the penultimate
ventral segment laterally bidentate............**Paradioxys** Mocs.
Eyes hairy ; postscutellum unarmed ; abdomen conical, in ♀ acuminate at
apex, in ♂ armed with porrect teeth or spines; maxillary palpi
3-jointed......................... ....**Cœlioxys** Latreille.

Family XI. PANURGIDÆ.

In this family I have placed all the Andrenoid bees, having but
two cubital cells in the front wings, and possibly this is the only
character that will hold them together, since, otherwise, characters
of mouth parts, tongue, labial palpi, etc., there is the greatest di-
versity in length and structure.   It is therefore a composite family,
or what the French would call *une famille de convenience.*

The family, as here defined, probably had its origin from three distinct sources : some of the genera are clearly recent developments
from *Andrena* and *Halictus*, others probably came from the Anthophoridæ, while still others have had a different or obscure origin.

The genera placed here are quite numerous, but I believe may be
easily recognizable by the characters made use of in the following
table :

Table of Genera.

Marginal cell at apex more or less remote from the costa, or broadly, obliquely truncate..5.

Marginal cell towards apex acuminate, attaining the costa.

Abdomen usually rather long; second joint of hind tarsi normal, inserted in the middle of the first........ ....2.

Abdomen short, subglobose, black, shining, with white fasciae at apex; second joint of hind tarsi angulate beneath, not inserted in the middle of the first; clypeus in ♂ yellow, the hind femora incrassated; antennae in both sexes filiform, longer than the head; tongue short; maxillary palpi 6-jointed, labial palpi 4-jointed...........**Macropis** Panzer.

2. Body, and more particularly the thorax, distinctly pubescent, the abdomen with white fasciae...3.

Body sparsely pubescent, the thorax rarely densely pubescent, the abdomen always glabrous, shining, not fasciate, although the anal segment is distinctly ciliate at apex.

Transverse median nervure interstitial, or very nearly.

Anal lobe in hind wings not longer than the submedian cell; head and thorax sparsely pubescent; abdomen in ♀ black, not at all fasciate; clypeus in ♂ black, the antennae longer than the thorax, the flagellar joints nodose beneath.........................**Halictoides** Nyl.

Anal lobe in hind wings distinctly longer than the submedian cell; head and thorax rather densely pubescent; abdomen in ♀ rufous or brownish, or at least reddish at apex of the segments; clypeus in ♂ yellow, the antennae not longer than the thorax, normal..**Parandrena** Robt.

Transverse median nervure not interstitial, joining the median vein *before* the basal nervure; antennae in ♂ not longer than the thorax, the flagellum simple; maxillary palpi 6-jointed, the joints subequal; labial palpi 4-jointed, the first joint the longest, about as long as joints 2 3 united, the third longer than joints 2 4.

Dufourea Lepel. = *Hemihalictus* Ckll.

3. Submedian cell usually shorter than the median, or never longer, the transverse median nervure joining the median vein *before* the origin of the basal nervure or interstitial with it...............................4.

Submedian cell a little longer than the median, the transverse median nervure joining the median vein *beyond* the origin of the basal nervure.

Thorax above with fulvous or ferruginous hairs; maxillary palpi 6-jointed; labial palpi 4-jointed.....**Biareolina** Dufour. = *Callandrena* Ckll.

4. Thorax above usually with a whitish or griseous pubescence, very rarely with a slight ochraceous tinge.

Labial palpi deformed, the basal joint long and quite different from the last; face in ♀ with blackish hairs; antennae in ♂ longer than the thorax, the apical joint attenuate from the middle.**Rhophites** Spin.

Labial palpi normal, all the joints being similar and nearly equal; face in ♀ with white hairs; antennae in ♂ as long as the thorax, the last joint acuminate at apex only.**Rhophitoides** Schenck = *Hesperapis* Ckll.

5. Marginal cell at apex more or less acuminate or narrowly rounded, not or rarely truncate, although sometimes appendiculate; mandibles dentate..11.

Marginal cell at apex truncate; mandibles at apex acute or narrowly rounded,
  not dentate
Front wings with two recurrent nervures.................................6.
Front wings with only one recurrent nervure, the second recurrent obliter-
    ated; cheeks with a tooth behind see below for characters of mouth
    parts .... **Perditella** Ckl.
* First cubital cell distinctly or much longer than the second; maxillary palpi
    6-jointed 7
First cubital cell equal to or not much longer than the second.
Maxillary palpi 4-4 jointed; abdomen back, smooth, shining bare above, with
    the sides and fifth segment fimbriate with white hairs; labial palpi
    4-jointed, the first joint almost as long as joints 2-4 united.
                                        Scrapter Lepel. Type = brat's Lepel.
Maxillary palpi 6-jointed; abdomen bare, with white bands; face and cly-
    peus white or with a white spot; second cubital cell receiving both
    recurrent nervures; submedian cell considerably shorter than the
    median cell ....**Camptopaeum** Spin.
" A series not almost entirely yellow.......................................
Species yellow or almost entirely yellow; labial palpi very long, 4-jointed the
    first joint fully twice as long as joints 2-4 united.
Frontal foveae very distinct, long linear, black; clypeus semi-circular at base
    the suture separating it from the face forms a semicircle; claws
    simple; pygidial plate almost trapezoidal ♀ ♂ unknown.
                                        Philoxanthus Ashm. Type P. braces Ckl.
Frontal foveae very small, represented by a rounded elevation black punctures,
    clypeus trapezoidal at base; claws cleft ♀ ♂ unknown to me.
                                                            Perditella Ckl.
* Abdomen never ochraceous or rufous and dark, mate with red, yellow or with blackish markings
    or bands; face usually yellow or marked with yellow or white; stigma
    well developed; recurrent nervures interstitial respectively with the
    first and second cubitus, or both are received by the second cubital cell.9
Abdomen never black, rufous or yellow, neither immaculate nor banded; stigma either
    large, well developed or poorly developed, subhem-ovate, both recur-
    rent nervures received by the second cubital cell, or the first is inter-
    stitial with the first transverse cubitus........................10.
" Marginal cell either short, much longer than the stigma, fully twice as long, and
    as long or nearly as long as the first discoidal cell.
    " the stigma well a little shorter than the median; stigma long, lanceolate,
    labial palpi 4-jointed, the first joint very long more times larger than
    joints 2-4 united, represented at base; head not wider than the thorax,
    claws cleft .... **Nomadopsis** Ashm.
    submedian cell smaller than the median; stigma short, flat, acute at
    apex; labial palpi 4-jointed, the first long and stout, but scarcely twice
    as long as the 4 united; head wider than the thorax; claws cleft.
                                        Spinoliella Ashm. Type Congr. cameroni Spin.
Marginal cell very short, shorter than the stigma or not longer, and always
    very much shorter than the first discoidal cell; labial palpi 4-jointed
    the first joint very long and usually somewhat thickened, fully twice
    as long, or even more than twice as long as joints 2-4 united; hind
    tibial spurs finely serrated

Claws in ♀ simple, in ♂ with the anterior and middle claws cleft, the hind claws simple **Cockerellia** Ashm. (Type *P. hyalina* Cr., *albipennis* Cr.).

Claws in both sexes cleft.......**Neoperdita** Ashm. (Type *P. zebrata* Cr.).

10. Marginal cell short, not longer than the stigma, usually shorter, the stigma large, well developed.

Head seen from in front rounded, or a little longer than wide, rarely wider than long; clypeus somewhat produced, truncate anteriorly, the labrum distinct, transverse; labial palpi 4-jointed, the first joint the longest, but rarely longer or much longer than joints 2-4 united; claws cleft.................**Perdita** Smith (Type *P. halictoides* Smith).

Marginal cell long, always much longer than the stigma.

Stigma rather small or narrow, lanceolate; head large, seen from in front usually much wider than long; tongue very long; labial palpi 4-jointed, the first joint very much longer than joints 2-4 united; claws cleft; hind tibial spurs finely serrate; antennæ short, scarcely as long as the width of the head........ ....**Macrotera** Smith.

Stigma broad, oblong-oval, or at least not lanceolate; head normal, as seen from in front rounded, not or very little wider than long; antennæ longer than the width of the head.

First cubital cell, along the cubitus, not greatly longer than the second.

Hind tibiæ and tarsi in ♀ with a long, dense pubescence; clypeus in ♂ black, with long hairs; legs black; labial palpi 4-jointed, the first joint very long, usually longer than 2-4 united; abdomen without hair fasciæ.......... ....**Panurgus** Latr.

Hind tibiæ and tarsi in ♀ with a short, rather sparse pubescence; clypeus in ♂ usually yellow or marked with yellow, hardly pubescent; legs black, varied with yellow; labial palpi 4-jointed, the first about as long as joints 2-4 united or somewhat longer.

Mesopleura bare or nearly; stigma in front wings very large, oval or elliptical; abdomen punctate, most frequently black, rarely rufous or rufous and black, the dorsal segments never with apical hair fasciæ.

Panurginus Nyl. = *Pseudopanurgus* Ckll.

Mesopleura with a whitish or cinereous pubescence; stigma in front wings narrower, not so large; abdomen black or æneous black, impunctured, or at most finely shagreened, the dorsal segments with hair fasciæ at apex, although sometimes interrupted.

. **Calliopsis** Smith (Type *C. andreniformis* Sm.).

First cubital cell, along the cubitus, about twice, or nearly, as long as the second; each cubital cell receiving a recurrent nervure; transverse median joining the median vein much before the origin of the basal.

Scapteroides Gribods.

11. Median and submedian cells of an unequal length, the transverse median nervure uniting with the median vein somewhat before the origin of the basal nervure; claws cleft.

Stigma large, broad, well developed, the marginal cell not short, longer than the first discoidal cell..12.

Stigma neither large nor broad, sublanceolate, the marginal cell short, obliquely truncate at apex and not quite as long as the first discoidal cell; second cubital cell narrowed one-half or more *above*; head seen from in front much broader than long; palpi as in *Macrotera* Smith abdomen in ♀ black, in ♂ red.

Macroteropsis Ashm. n. g. (Type *Perdita latior* Ckll.).

12. Cubital cells, along the cubitus, very unequal, the first about thrice as long as the second, the latter quadrate or nearly ; first recurrent nervure received by the first cubital cell near the apex, the second recurrent by the second cubital cell also near the apex ; submedian cell much shorter than the median ; head not wider than the thorax, but seen from in front oblong, about twice as long as wide, the eyes fully thrice as long as wide or more ; antennæ clavate ; mandibles bidentate at apex ; maxillary palpi 6-jointed ; labial palpi 4-jointed, the first joint the longest. . . . .**Hylæosoma** Ashm.

Cubital cells, along the cubitus, equal or nearly, the first usually a little the longer ; transverse median nervure interstitial with the basal nervure ; head normal.

Second cubital cell narrowed fully one-half or more above ; head and thorax pubescent, bare or nearly on vertex and metathorax ; flocculus on hind tibiæ and tarsi normal ; abdomen rufous, the dorsal segments 2-4 delicately but not sharply depressed at apex ; clypeus with a transverse impression anteriorly. South America (Argentine).

Perditomorpha Ashm. n. g. (*P. Brunerii* Ashm.).

Second cubital cell narrowed one-third above ; thorax clothed with a dense pubescence ; flocculus on hind tibiæ and tarsi long, dense ; abdomen black, pubescent and fasciated ; maxillary palpi 6-jointed ; labial palpi 4-jointed, subequal, the joints enlarged at apex. . .**Dasypoda** Latr.

Family XII. ANDRENIDÆ.

(The Acute-tongued Burrowing Bees).

This family is of great extent and very numerous in species, which are closely allied and extremely difficult to separate. They are found widely distributed over all parts of the world and in every clime, climate apparently having little effect upon the distribution of certain of the genera and species. Most of them, but not all, have short acute tongues, and are found in late Spring and Mid-Summer making their burrows in clay banks and in hard, compact soils of various kinds, hence the name given to them above seems most appropriate.

Some Andrenids in color and on account of the dense pubescence on the head and thorax bear a striking resemblance to some of the Anthophorids, and may be easily confused with them. The labrum is, however, never large and free, although more or less distinctly visible, while the joints of the labial palpi are cylindrical, all much alike in general appearance, the first two joints never compressed, valvate as in the Apidæ, Bombidæ, Anthophoridæ, etc.

A little care at first, as to the general habitus of the venation, mouth parts, etc., will soon enable the student to separate at a glance an Andrenid from other bees.

The three cubital cells in the front wings, as well as usually the much shorter tongue, separate them from the Panurgidæ, which have only two cubital cells. Some Panurgid genera, however, have also short tongues and are in no way to be distinguished from the Andrenids, except in having only two submarginal cells. They are evidently recent offshoots from an Andrenid stock.

The family may be separated into three subfamilies as follows :

Table of Subfamilies.

First branch of basal nervure always strongly curved or bent inwardly, forming a segment of a circle ; epimera of mesothorax well separated, distinct . . . . .2.

First branch of basal nervure straight or very nearly, never strongly curved inwardly ; epimera of mesothorax not or scarcely separated ; hind tibiæ in both sexes with a knee plate...Subfamily I. ANDRENINÆ.

2. Hind femora and tibiæ in ♀ with a distinct flocculus or scopa, the ventral scopa sparse, but distinct; fifth abdominal segment always *with* a rima. . . . . Subfamily II. HALICTINÆ.

Hind femora and tibiæ in ♀ without a distinct flocculus or scopa, when present very sparse and thin, and scarcely noticeable; fifth abdominal segment *without* a rima; venter bare. . . . . Subfamily III SPHECODINÆ.

Subfamily I. ANDRENINÆ.

The bees belonging to this group are exceedingly numerous and common, appearing very early in the Spring and lasting all during the Summer months, being most active in nest building during the months of June and July. Their nests are made in burrows or tunnels in the ground or in hard clay banks, preferably the latter. The tongue in all the genera is short and triangularly pointed, the labial palpi 4-jointed, the maxillary palpi usually 6-jointed, while the basal nervure is straight or nearly, never curved or strongly bent inwardly, as in the Halictinæ and the Sphecodinæ ; the scopa on the hind legs, in ♀ , being always distinct, well developed.

Table of Genera

Front wings with three cubital cells.

Marginal cell at apex acuminate or narrowly rounded, never truncate. . . .4

Marginal cell at apex obliquely truncate and usually with a short appendage (or stump of a vein) from the lower angle. . . . . .2.

2. First cubital cell, along the cubitus, as long as the second and third united or nearly. . . . . .3.

First cubital cell, along the cubitus, about as long as the third, the latter narrowed one-half above along the radius, and receiving the second recurrent nervure far beyond its middle ; second cubital cell quadrate, receiving the first recurrent nervure beyond its middle ; head very broad, wider than the thorax, seen from in front wider than long; labrum distinct, but strongly inflexed, twice as wide as long; mandibles simple, acute at apex ; labial palpi 4-jointed, slender ; middle tibial spur and one of the hind tibial spurs finely serrated ; abdomen broadly oval, wider than thorax and marked with yellow.

Liphanthus Reed.

Second and third cubital cells each receiving a recurrent nervure beyond the middle ; labial palpi 4-jointed ; maxillary palpi 6-jointed ; middle and hind tibial spurs pectinate or finely serrated.

Head distinctly and sometimes much wider than the thorax, marked with yellow or white, the superorbital foveæ long, linear ; abdomen not elongate, always banded or marked with yellow or white, the dorsal segments broadly depressed at apex and differently sculptured from the anterior portion, the first segment with a median grooved line at base ; first joint of labial palpi the longest and stoutest, joints 2–4 subequal.

Psæuythia Gerst.

Head not wider than the thorax, although marked with yellow or white anteriorly, the superorbital foveæ long, linear ; abdomen much elongate, black, neither banded nor maculated ; first joint of labial palpi the longest and stoutest joint ; mandibles acute.

Protandrena Ckll.

Second cubital cell receiving the first recurrent nervure near its base, the second recurrent interstitial with the third transverse cubitus ; labial palpi 4-jointed, the joints subequal, short, stout and cylindrical, but gradually becoming slenderer ; abdomen oblong-oval, fasciate, with white hair bands...**Apista** Smith.

4. First cubital cell, along the radius, usually much longer than the third.....6.

First cubital cell, along the cubitus, not or scarcely longer than the third, equal or nearly, or a little shorter, the second small, quadrate or transverse-quadrate, the third narrowed above............... ....5.

5. Tegulæ normal ; axillæ never acute behind ; abdomen neither banded nor fasciate, the dorsal segments 1–4 depressed at apical margin ; all tibial spurs pectinate ; maxillary palpi 6-jointed ; labial palpi 4-jointed, the first as long as joints 2–4 united and also stouter ; mandibles obtusely bidentate ; hind femora in ♂ not greatly swollen, their tibiæ, however, dilated towards apex and produced into a strong spine beneath.

Epinomia Ashm. n. g. (Type *N. persimilis* Ckll.).

Tegulæ very large ; axillæ sometimes acute or toothed posteriorly ; hind femora in ♂ abnormally swollen or enlarged.

♀ with a long, distinct, middle tibial spur ; ♂ with the apical joint of antennæ simple, not compressed, spoon shaped.

Abdomen with the dorsal segments apically depressed and fasciate with whitish hair bands ; ♂ antennæ normal, not tapering to a point at apex...................................... ....**Nomia** Latreille.

Abdomen with the dorsal segments 2–5 apically banded with smooth, greenish, yellow or whitish non-pubescent bands ; ♂ antennæ tapering off to a point at apex......................**Paranomia** Friese.

♀ without a middle tibial spur ; ♂ with the apical joint of antennæ compressed, oblong, excavated beneath or spoon shaped.

Monia West. = *Ennomia* Cress.

6. Second cubital cell, along the cubitus, much smaller than the third, rarely much longer than half its length or even shorter........................7.

Second cubital cell, along the cubitus, as long as the third or even a little longer.

Second and third cubital cells equal, the former nearly quadrate, very slightly narrowed above, and receiving the first recurrent nervure at the middle, the third cubital cell receiving the second recurrent a little beyond its middle; maxillary palpi 6-jointed, the first joint the longest, the following gradually decreasing in length; labial palpi 4-jointed, the first joint the longest, the following shorter, subequal.

Callomelitta Smith.

Second cubital cell a little longer than the third, narrowed above and receiving the first recurrent nervure at its middle; the third cubital cell also receives the second recurrent at its middle; maxillary palpi 6-jointed, the first joint longer and stouter than the others, the following successively becoming shorter and slenderer; antennæ clavate, the scape short and stout, one-third shorter than the third joint.

Gastropsis Smith = Œstropsis Smith.

7. Submedian vein (or anal vein of some authors) in hind wings distinct to the hind margin; hind trochanters in ♀ with a polleniferous flocculus; antennæ in ♂ not truncate at apex...................................8.

Submedian vein in hind wings abbreviated, not extending to the hind margin; hind trochanters in ♀ without a polleniferous flocculus; antennæ in ♂ sometimes truncate at apex.

Head normal, not wider than the thorax; abdomen truncate at base; antennæ in ♂ truncate at apex, the joints subserrate or crenulate beneath.

Melitta Kirby = Cilissa Leach.

Head very large, transverse, much wider than the thorax; abdomen rounded at base; antennæ in ♂ filiform, neither truncate at apex nor crenulate beneath.......**Sphecophala** Sauss. (Type S. philanthoides Sauss.).

8. Stigma very minute.

Second cubital cell very short, wider (higher) than long, receiving the first recurrent nervure a little before the middle, the second recurrent being interstitial or nearly with the third transverse cubitus, head transverse, not so wide as the thorax; ocelli in a triangle; tongue very short, densely pubescent, the paraglossæ elongate, plumose; labial palpi 4-jointed, short and stout; maxillary palpi 6-jointed.

Ptiloglossa Smith.

Stigma distinct, not small.

Marginal cell pointed at apex, its extreme apex attaining the costa; submedian cell not quite as long as the median; second cubital cell quadrate, receiving the first recurrent nervure at the middle, the third cubital cell receiving the second recurrent nervure at its apical third; abdomen above bare, the dorsal segments 2–4 delicately, but not sharply depressed at apex; pygidial plate triangular, with a median ridge; middle tibial spurs long, finely serrated; maxillary palpi 6-jointed; labial palpi 4-jointed; superorbital foveæ wanting.

Micrandrena Ashm. n. g. (Type M. pacifica Ashm.).

Marginal cell more or less narrowly rounded at apex, its extreme tip not
attaining the costa; submedian cell fully as long as the median, the
transverse median nervure interstitial.

Second cubital cell a little longer than wide, receiving the first recurrent
nervure a little beyond the middle, rarely at the middle: the third
cubital cell receiving the second recurrent far beyond the middle;
abdomen most frequently fasciate or subfasciate with hair bands,
more rarely bare; pygidial plate subtriangular, usually subtruncate
at apex; maxillary palpi 6-jointed, labial palpi 4-jointed; superorbital
fovea very large, broad, represented by broad depressions along the
inner upper orbits.....................................**Andrena** Latr.

Second cubital cell receiving the first recurrent nervure at the middle, the
second recurrent received by the third cubital cell at its middle; middle tibial spur and the longer spur of hind tibiæ, which is very long
and bent, serrated; maxillary palpi 6-jointed, labial palpi 4-jointed.
Stenotritus Smith.

Second cubital cell wider (higher) than long, narrowed above; along the
cubitus about half as long as the first cubital cell, and reaching the
first recurrent nervure near the apex; abdomen fasciate; hind tibiæ
in ♂, strongly curved and angulately dilated at apex beneath.
Ancyla Lepel. = *Pristotrichia* Radoszk.

Subfamily II. HALICTINÆ.

The bees belonging to this subfamily agree with the Andreninæ
in all essential characters, but the basal nervure in the front wings
is always strongly curved or bent inwardly towards base of the wing,
the epimera of the mesothorax well separated, distinct, while the
apical dorsal segment in the ♀ always has a distinct rima, or median
grooved furrow on its disk.

The tongue may be either long or short. The rima on the last
dorsal segment, and the distinct scopa or flocculus on hind legs in ♀
must be depended upon to separate these bees from the Sphecodinæ.

Eleven genera are known, recognizable by the aid of the following
table :

Table of Genera.

First cubital cell, along the cubitus, distinctly longer than the third, or as long
as the second and third united; stigma well developed; transverse
median nervure never not angulated...............................2.

First cubital cell, along the cubitus, as long or very nearly as long as the third :
stigma not well developed; transverse median nervure angulated;
maxillary palpi 6-jointed.

Abdomen black, densely pubescent; antennæ in ♀ short, clavate, in ♂
involute at apex, the last joint triangular; labial palpi 4-jointed.
Systropha Latreille.

Abdomen black and rufous, almost bare; antennæ filiform, in ♂ long; mentum and tongue long, slender; tibiæ dilated at lower apical angle.

Trichchostoma Sauss.

First cubital cell, along the cubitus, shorter than the third; stigma well developed; transverse median nervure straight, interstitial with the basal nervure; scutellum spined; mandibles tridentate; maxillary palpi 6-jointed.

Abdomen very smooth, shining; head and thorax clothed with fulvous hairs; labial palpi 4-jointed, the first joint about as long as joints 2-4 united......................................**Mellitiden** Guerin.

2. Abdomen normal, not petiolate-clavate........   ....3.

Abdomen petiolate-clavate; second cubital cell half the length of the third, slightly narrowed above; scape of antennæ two-thirds the length of the flagellum, the latter subclavate; maxillary palpi 6-jointed, the three basal joints short, stout, clavate, joints 4-6 slender, slightly thickened towards apex; labial palpi 4-jointed, the first subclavate, as long as 2-3 united, the two latter short, stout and clavate, the last joint slender, filiform........................**Corynura** Spinola.

3. Third cubital cell receiving only one recurrent nervure—the second, the first recurrent received by the second cubital cell beyond its middle, or before the first transverse cubitus (or very exceptionally interstitial with this nervure)...4.

Third cubital cell receiving both recurrent nervures, or the first recurrent is interstitial with the second transverse cubitus; rarely is the second recurrent interstitial with the third transverse cubitus; maxillary palpi 6-jointed; labial palpi 4-jointed.

Tongue elongate, lanceolate, the paraglossæ long; first joint of labial palpi longer than 2-3 united, the second a little shorter than the third, the third obconical, the last slender, cylindrical; species metallic-blue, blue-green to green........**Augochlora** Smith = *Oxystoglossa* Sm.

Tongue short, triangular, the paraglossæ not long; first joint of labial palpi about as long as 2-3 united; first recurrent nervure interstitial with the second transverse cubitus; non-metallic species, or at most æneous on the head and thorax; clypeus in ♂ anteriorly always margined with yellow...............................**Parasphecodes** Smith.

4. First cubital cell, along the cubitus, as long or somewhat longer than the second and third united, the second quadrate, not or only slightly longer than wide, or much wider (higher) than long, and often considerably narrowed above.....................................5.

First cubital cell, along the cubitus, not longer than the second and third united, usually distinctly shorter, the second quadrate, not wider than long.

Head and thorax metallic, green, blue or blue-green; abdomen in ♀ metallic, black or yellow, in ♂ usually yellow, with black bands, the hind femora in this sex being much swollen, and most frequently with a subapical tooth beneath..................**Agapostemon** Smith.

5. Second recurrent nervure not interstitial with the third transverse cubitus, the second cubital cell not longer than wide, usually wider (or higher) than long; scutellum normal, tongue short, triangular or subtriangular, or in outline cone shaped.......................................6.

Second recurrent nervure interstitial with the third transverse cubitus; second
   cubital cell quadrate, the third a little longer than wide, slightly narrowed above; tongue very long, spiculiform; labial palpi 4-jointed, the
   first joint nearly as long as joints 2–4 united. **Megaloptera** Smith.
6. Subdiscoidal nervure in hind wings distinct, well developed; temples broad,
   or at least never very narrow, sometimes in *Halictus* with a tooth
   below in ♀: labial palpi 4-jointed.
   Ocelli very large, the lateral ocelli almost touching the eye margin.
                 Sphecodogastra Ashm. n. g. (Type *Parasp. texana* Cr.)
   Ocelli normal, the lateral ocelli very distant from the eye margin.
                                                 Halictus Latreille.
Subdiscoidal nervure in hind wings wanting or subobsolete; temples very narrow, flattened; labial palpi 2-jointed; abdomen maculate with yellow.
                 Nomiodes Schenck = *Lucasius* Dours.

Subfamily III. Sphecodinae.

This subfamily is scarcely distinguishable from the subfamily
Halictinæ, agreeing with it in all essential characters, the only differ-
ences noticeable, and which may be depended upon to separate the
species from those in the latter, are that in the ♀ the hind coxæ and
femora are *without* a flocculus, the pollen-brush on their tibiæ and
tarsi, although denser on the tarsi, is sparse and thin or nearly
wanting, the abdomen being without a distinct ventral scopa, and
the last dorsal segment is *without a distinct rima.*

I know of no good character to separate the males from those in
the Halictinæ. The head, however, viewed from in front, is a little
wider than long, the clypeus not being produced anteriorly, while
the metathorax, as a rule, is much more coarsely sculptured than in
the Halictinæ, being usually coarsely rugoso-punctate.

The genus *Parasphecodes* Smith, as I have identified it, from a
specimen in the National Museum from Australia, is closely allied
to *Halictus* and belongs to the Halictinæ. *Parasphecodes texana*
Cress. has been wrongly referred to this genus and forms a new
genus in the same subfamily.

Only three genera seem to fall into this group as here defined,
and these may be distinguished by the use of the following table:

Didonia Gribodo is, however, placed here without an examination of specimens.

Table of Genera.

First cubital cell, along the cubitus, distinctly longer than the second and third
   united; stigma well developed; subdiscoidal nervure originating at

or a little above the middle of the second transverse median nervure
(discoidal nervure); second cubital cell very short, fully twice as wide
(or high) as long, first recurrent nervure interstitial with the second
transverse cubitus, the second interstitial or nearly with the third
transverse cubitus, or received by the hind cubital cell at or beyond
its apical third.

Abdomen coarsely punctured, segments 1-3 with a distinct rim or elevated
margin at apex; tongue very short, triangular; maxillary palpi
6-jointed; labial palpi 4-jointed, the first a little the longest joint,
joints 2-4 subequal.....................................**Temnosoma** Smith.

Abdomen not coarsely punctured, segments 1-3 not rimmed at apex; tongue
elongate, linear, subspiculiform; labial palpi 4-jointed, elongate, almost
attaining to the apex of tongue, the first longer than the two following
united, the last joint the smallest..............**Didonia** Gribodo.

First cubital cell, along the cubitus, not or scarcely longer than the second and
third united; stigma well developed; subdiscoidal nervure originating
below the middle of the second transverse median nervure; second
cubital cell short, wider than long; first recurrent nervure received
by the second cubital cell a little beyond the middle, the second recurrent received by the third near its apex; maxillary palpi 6-jointed;
labial palpi 4-jointed, the first nearly as long as joints 2-3 united.
Sphecodes Latr.

Family XIII. COLLETIDÆ.

(The Obtuse-tongued Burrowing Bees).

This family, with the next, the Prosopidæ, constitute Westwood's
group or section *obtusilingues*.

Bingham has associated both together on account of similarity in
the mouth parts, under the name Colletidæ; but there is a wide
difference in the habits of the species composing the two families, as
here defined, and also in their external structural characters.

Those I have placed in this family are clothed with a more or less
dense pubescence on the head and the thorax, while in the Prosopidæ they are bare or nearly, the legs are also more densely bubes-
cent, the hind tibiæ and tarsi in ♀ always with a distinct pollen-
brush, while the front wings have three cubital cells, whereas the
Prosopidæ only have two.

The economy of the species are similar to those in the family
Andrenidæ, since they construct their burrows in hard clay soil, in
clay banks or in the interstices of stone walls, etc. The Prosopidæ,
on the contrary, burrow in the stems or twigs of bramble and
various shrubs.

Ten genera seem to fall into this family and are tabulated below:

Table of Genera.

First cubital cell, along the cubitus, fully as long as the second and third united or nearly...2.

First cubital cell, along the cubitus, shorter than the second and third united.

Second cubital cell longer than wide, receiving the first recurrent nervure at or a little beyond the middle; third cubital cell narrowed at least one-half above.

First joint of maxillary palpi a little longer and stouter than the second, joints 3-5 subequal, a little longer than thick, the last joint longer than the fifth: joint 1 of labial palpi long, about as long as joints 2-3 united, the last joint shorter than the second, but longer than the third; paraglossæ dilated and rounded at their apices...**Anthoglossa** Smith.

First joint of maxillary palpi a little longer, but scarcely thicker than the second, the following joints very gradually shortening, all more than thrice as long as thick; labial palpi 4-jointed, the first two joints stout, the first much the longer, longer than 3-4 united, the latter being slender and subequal, a little shorter than the second.

Diphaglossa Spinola.

Second and third cubital cells equal, the second subquadrate, only slightly narrowed above; first recurrent nervure interstitial or nearly.

First joint of maxillary palpi a little the longest joint and stoutest, the following subequal; labial palpi 4-jointed, the first about as long as 2-3 united, the third the shortest..........**Caupolicana** Spinola.

2. Second and third cubital cells, along the cubitus, equal or nearly............3.

Second and third cubital cells, along the cubitus, scarcely longer than the first, the second often wider (higher) than long, very much shorter than the third; marginal cell narrowly obliquely truncate at apex.

First recurrent nervure interstitial with the first transverse cubitus, or received by the first cubital cell just before this vein; third cubital cell receiving the second recurrent nervure towards it apex; submedian cell much shorter than the median..........**Megacilissa** Smith.

First and second recurrent nervures respectively interstitial with the first and second transverse cubital nervures; labial palpi 4-jointed, the joints successively decreasing in length; antennæ subclavate, the third joint only a little longer than the fourth, the following joints increasing in length to the apical joint.......**Mudrosoma** Smith.

3. Second recurrent nervure interstitial with the third transverse cubitus; inner spur of hind tibiæ pectinate or combed; joints 1-2 of maxillary palpi stout, subequal, the following joints slender, more than thrice longer than thick; first joint of labial palpi elongate, as long as joints 2-3 united; joints 2-4 also long............**Lamprocolletes** Smith.

Second recurrent nervure not interstitial, but received by the third cubital cell beyond its middle.

Stigma not well developed, or subobsolete... ....4.

Stigma well developed, although not large.

Second recurrent nervure sinuate or somewhat S shaped; hind tibiæ without knee plate: maxillary palpi 6-jointed, the first joint hardly longer than 2-3 united, joints 2-5 scarcely longer than thick, the last a little shorter than the first; labial palpi short, 4-jointed...**Colletes** Latr.

4. Second cubital cell very short, much wider (higher) than long, and, along the
cubitus, only about one-third the length of the third, receiving the
first recurrent a little before the middle; third cubital cell nearly as
long *along the radius* as along the cubitus; joints 2-4 of maxillary palpi
longer than thick; first joint of labial palpi long, longer than joints
2-3 united...**Paracolletes** Smith.

Second cubital cell a little longer than wide and more than half the length of
the third, the first recurrent nervure received by it at the middle;
third cubital cell, along the radius, only about one-third as long as
along the cubitus; joints 2-4 of maxillary palpi three or more times
longer than thick; first joint of labial palpi very elongate.

Leioproctus Smith.

Second cubital cell equal narrowed on each side towards the marginal and
receiving the first recurrent nervure at its middle; joints 1-2 of maxillary palpi rather long, subequal, both of which are about as long as
joints 2-3 united; first joint of labial palpi long, about as long as 2 3
united, the last joint shorter than the third; paraglossæ broad and
rounded at apex.........................**Dasycolletes** Smith.

Family XIV. PROSOPIDÆ.

(The Obtuse-tongued Carpenter Bees).

This is a small but distinct group or family, at one time supposed
to be parasitic, the species agreeing in their mouth parts with the
Colletidæ (and many of the wasps), but are readily distinguished
from them by having *only two cubital cells* in the front wings, the
non-pubescent body, and by the hind tibiæ being *without a distinct
pollen brush.*

They are known to burrow into the twigs of bramble, elder and
other shrubs, in which, after extracting the pith, they construct
their cells—filled with pollen and honey.

Table of Genera.

Marginal cell at apex acuminate, or narrowly rounded, never truncate; mandibles bidentate; body bare or nearly................................2.
Marginal cell at apex somewhat obliquely truncate, with an appendage; body
pubescent.

Stigma sublanceolate; first and second cubital cells subequal, the first slightly
the longer, the second receiving both recurrent nervures; maxillary
palpi 6-jointed, the first joint as long as joints 2-3 united, the following
subequal, labial palpi 4-jointed, the first long and stout, longer than
the following united......................**Pasiphæ** Spinola.

2. First cubital cell twice as long as the second, or very nearly..............4.
First cubital cell equal to the second, or somewhat smaller, or the second is
shorter than the first.

First recurrent nervure received by the first cubital cell just before the
first transverse cubitus, the first cubital cell much longer than the
second. ....3.

Both recurrent nervures received by the second cubital cell, or the second recurrent is interstitial with the second transverse cubitus.

Second cubital cell shorter than the first; labial palpi rather short, the first joint shorter than joints 2-3 united; tongue only slightly emarginate at apex...............................**Euryglossa** Smith.

Second cubital cell as long as the first or very nearly; labial palpi elongate, the first joint as long as joints 2-5 united; tongue deeply triangularly emarginate.........................**Hylæoides** Smith.

3. Head seen from in front a little longer than wide, slightly narrowed below; frontal fovere distinctly long, linear; maxillary palpi 6-jointed, the joints short, subequal; labial palpi 4-jointed.......**Prosopis** Fabr.

4. Both recurrent nervures received by the second cubital cell, or the first recurrent is interstitial with the first transverse cubitus; sometimes the second recurrent is interstitial with the second transverse cubitus (very exceptionally does the first recurrent joint the first cubital cell just before the first transverse cubitus); maxillary palpi 6-jointed; labial palpi 4-jointed.

Stigma very small, inconspicuous or poorly developed, the transverse median nervure not interstitial...6.

Stigma distinct, well developed, the transverse median nervure interstitial or nearly with the basal nervure.....................................5.

5. Maxillary joints short, subequal; labial palpi 4-jointed, short, the first two a little stouter and longer than the last two; head seen from in front a little longer than wide, slightly narrowed below, the frontal foveæ distinct, long, linear.............................**Prosopis** Fabr.

Basal three joints of maxillary palpi stout, subequal, joints 3-6 much slenderer and clavate; basal joint of labial palpi somewhat longer than the second, joints 2-3 subequal, the last cylindrical.

                                 Stilpnosoma Smith.

6. First joint of maxillary palpi the shortest and stoutest joint, the third slightly the longest, joints 4-5 subequal; first joint of labial palpi the longest, joints 2-4 subequal...................   ....**Meroglossa** Smith.

ERRATA.

Page 55, line 32, for golden *read* pollen.

"   58,   "   20, for cleft *read* tuft.

"   61,   "   5 from bottom, for *graga* read *graja*.

"   63,   "   14, for **Xenogloss** read **Xenoglossa.**

"   66,   "   1, *insert* abdomen *after* third.

"   68,   "   4, for 16 *read* 15.

"   72,   "   13, for mediam *read* median.

"   73,   "   11, for Jarine *read* Jurine.

INDEX.